Haptic Rendering for Simulation of Fine Manipulation

Dangxiao Wang · Jing Xiao
Yuru Zhang

Haptic Rendering
for Simulation
of Fine Manipulation

 Springer

Dangxiao Wang
Yuru Zhang
Beihang University
Beijing
China

Jing Xiao
University of North Carolina at Charlotte
Charlotte, NC
USA

Additional material to this book can be downloaded from http://extras.springer.com/

ISBN 978-3-662-44948-6   ISBN 978-3-662-44949-3   (eBook)
DOI 10.1007/978-3-662-44949-3

Library of Congress Control Number: 2014950648

Springer Heidelberg New York Dordrecht London

Printed on acid-free paper

Springer is part of Springer Science+Business Media (www.springer.com)

Foreword

Extending the frontier of visual computing, touch-enabled interaction, such as haptic rendering, is an alternative form of human–computer interfaces and information presentation, in addition to graphical and auditory display. Haptic rendering offers a unique, bidirectional communication between humans and interactive systems through tactile sensory cues, providing a more natural and intuitive interface. Following emerging development and recent advances in haptic rendering, this book by Wang et al. presents the latest progress on six-degree-of-freedom (6-DoF) haptic display for fine manipulation.

Most physical objects in the real world require fine manipulation and control to handle them. Our hands grasp and exert forces to move objects and perform dexterous manipulation in assembly, inspection, routine maintenance, surgery, and many other complex operations. To perform similar tasks using haptic devices introduces significant computational challenges. This book describes several key concepts and suggests new approaches for performing 6-DoF fine manipulation of rigid and deformable objects with complex geometry using haptic devices. The seamless blending of design, algorithms, and application provides an excellent overview of this important topic for both practitioners and researchers working in fine manipulation and haptic rendering.

Chapel Hill, NC, USA, September 2014 Ming C. Lin

Preface

Haptic rendering is the process of computing and generating forces in response to user interactions with virtual objects. Six-degrees-of-freedom (6-DoF) haptic rendering simulates both forces and torques during physical interactions between a virtual tool moved by a human user via a haptic device and the objects in a virtual environment. Both the tool and the objects can possess complex shapes. Forces and torques exerted on the virtual tool are computed and displayed to the haptic device. In this book, we introduce the latest progress on 6-DoF haptic rendering for fine manipulation.

Fine manipulation is pervasive in the real environment, such as in surgical operation, mechanical assembly, inspection, and maintenance tasks. 6-DoF haptic simulation of fine manipulation is a difficult problem because of multi-region contacts, narrow space constraints, and requirements to preserve the appearance and tactile sensation of fine-features. Compared to the fields of computer graphics and speech synthesis, haptic rendering is still a young research field. High performance rendering methods are needed to simulate physical responses in various fine manipulation scenarios, such as the nonlinear behavior of human tissue during surgical process, or the contact forces from multi-region contacts involving fine geometric features during assembly of complex structures such as an airplane.

In this book, we introduce a systematic, constraint-based approach for 6-DoF haptic rendering that is particularly suitable for fine manipulation. We also introduce its application to virtual training for dental operations. We provide programs, 3D models, and videos for download. All these materials are available online from the book website http://extras.springer.com/.

The structure of this book is organized as follows. In Chap. 1, we provide a brief introduction to the history of haptic rendering and a literature survey on 6-DoF haptic rendering. We next introduce the concept of fine manipulation and associated computational challenges for haptic rendering. In Chap. 2, we introduce a configuration-based optimization approach based on modeling non-penetration constraints by sphere-trees, which can realize stable and responsive haptic rendering between rigid objects with 1kHz. In Chap. 3, we extend the approach to simulating fine geometric features, such as sharp edges. In Chap. 4, we extend the approach to simulating deformable objects and hybrid contacts, which are characterized by simultaneous interactions between a tool with both rigid and deformable objects. In Chap. 5, we introduce a measurement-based approach for achieving objective and quantified evaluation of the fidelity of 6-DoF haptic rendering, which is not only applicable to the introduced approach in this book but also to other existing haptic rendering approaches. In order to provide an intuitive application and validation of the introduced haptic rendering approach, in Chap. 6, we describe the design of a dental simulator that validates the approach on 6-DoF haptic rendering and provides the functions and user evaluation results of the system. Finally, in Chap. 7, we summarize the book by discussing open challenges and possible research topics in this exciting field.

Target readers of the book include but are not limited to:

- Researchers and scholars in the field of haptics;
- Software developers and engineers for developing haptic applications, including virtual surgery, computer games, virtual product prototyping, upper limb rehabilitations, etc.;
- Researchers and developers in dental education and professional training;
- Graduate students in related majors, such as computer graphics, animation, virtual reality, and robotics.

We are grateful for the support of the Natural Science Foundation of China (under grant No. 50275003, 50575011, 61170187 and 61190125) to our research on this topic. We also thank our collaborators, Prof. Peijun Lü, Prof. Yong Wang, and Prof. Jianxia Hou in the School and Hospital of Stomatology, Peking University, Beijing, China. They provided valuable dental knowledge and support for carrying out user evaluation for the iDental surgical simulator. We thank our current and former graduate students, Dr. Ge Yu, Dr. Jun Wu, Ms. Yu Wang, Ms. Renge Zhou, Ms. Wanlin Zhou, Mr. Xin Zhang, Mr. Zhixiang Wang, Mr. Hui Zhao, Ms. Shuai Liu, Ms. Youjiao Shi, Mr. Zhongyuan Chen, Mr. Hao Tong and Mr. Xiaohan Zhao. Their hard work provided the foundation for this book. Last but not the least, we thank Dr. Lanlan Chang for her support and valuable help in preparing the book, and Ms. Jane Li for organizing the editing process of the book. Their input was very helpful for improving the manuscript.

July 2014   Dangxiao Wang
                                                                          Jing Xiao
                                                                         Yuru Zhang

Contents

Chapter 1
Introduction

In this chapter, we provide an overview of haptic interaction systems and define haptic rendering, highlighting the difference between 3-DoF and 6-DoF haptic rendering. We also introduce the three main approaches of haptic rendering. Finally, we discuss features of fine manipulation and the associated computational challenges to haptic rendering.

1.1 Haptic Interaction Systems

We first provide a brief introduction to human haptic perception and manipulation. Then, we describe a classification of haptic interfaces and components of a haptic interaction system. Finally, we introduce several example applications.

1.1.1 Human Haptic Perception and Manipulation

The word *haptic*, derived from the Greek word, "*haptesthai*," means "related to the sense of touch." The sense of touch can be divided into cutaneous, kinesthetic, and haptic systems, based on the underlying neural inputs (Klatzky and Lederman 2003). The cutaneous system employs receptors embedded in the skin, while the kinesthetic system employs receptors located in muscles, tendons, and joints. The haptic sensory system employs both cutaneous and kinesthetic receptors, but it differs in that it is associated with an active procedure controlled by body motion (Lin and Otaduy 2008).

In our daily life, humans interact with the real world through five sensory channels, namely, visual, auditory, haptic, olfactory, and gustatory. Among all the channels, the haptic channel plays fundamental functions to enable manipulation and active exploration of the physical world that the other senses cannot. Some features can only be perceived by haptics, including the hardness, roughness (friction or smooth), texture, weight, and shape. (Klatzky and Lederman 2003).

© Springer-Verlag Berlin Heidelberg 2014
D. Wang et al., *Haptic Rendering for Simulation of Fine Manipulation*,
DOI 10.1007/978-3-662-44949-3_1

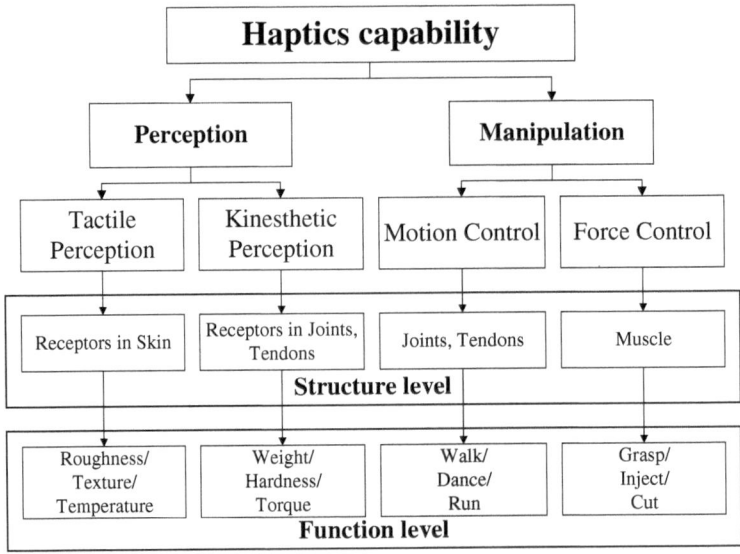

Fig. 1.1 Classification of human haptic perception and manipulation capability

Among all senses, the human haptic system is the only key sensory channel that provides the unique, bidirectional communication between humans and their physical surroundings. Unlike visual and auditory sensory channels, haptic interaction consists of both perception and manipulation (Fig. 1.1). The haptic channel is necessary to assist humans in fine manipulations such as using chopsticks to pick a peanut or using five fingers to grasp a boiled egg without the shell.

Skin is our primary tactile sensory organ, which includes three main functional categories of tactile receptors, all located in different layers of the skin. Thermoreceptors recognize changes in skin temperature; mechanoreceptors (of which there are a half dozen) sense pressure, vibration, and slip; nocioreceptors sense pain. To encompass sensations ranging from pin-pricks to broad steady pressure, they vary in their mechanical principle of action, skin depth, and response speed (MacLean 2008; Jones 1998). Merkel Receptors respond to pressure at about 0–10 Hz, Meissner Corpuscles respond to "taps" within 3–50 Hz, Ruffini Cylinders respond to stretching of skin or movement of joints at 0–10 Hz, and Pacinian Corpuscles respond to rapid vibration within 100–500 Hz. The "sweet spot" for vibrotactile sensitivity is considered to be 250 Hz (Shimoga 1992).

Psychophysics is an important branch of haptic interaction technology. It studies human perception capability on various physical properties, including weight, force, torque, stiffness, and length. There are two concepts called detection threshold and discrimination threshold. The just noticeable difference (JND), expressed as a percentage, is a common measure of both tactile and proprioceptive sensory resolution. A lot of work has been done to measure human's JND on different properties. Readers can refer to several prominent reviews on these topics

(Jones and Tan 2013; Leadman and Klatzky 1997; Klatzky and Lederman 1999, 2003; Tan et al. 2006).

Proprioception (also referred to as kinesthesia) is how we get information from most force-feedback displays. Its receptors are generally embedded in our muscle fibers and joints, although sometimes skin stretching also gives cues. There are two primary types of muscle mechanoreceptors. "Force sensors" (Golgi tendon organs) measure force via localized tension and are located serially between muscles/ tendons. "Position and motion sensors" (muscle spindles) are located in parallel among muscle fibers and are excited by changes in muscle length (e.g., active and passive stretching). Each of these sensor types plays a special role in motor control.

Human's haptic manipulation capability refers to motor control skill, which can be divided with regard to the motion or force control, joint type, number of simultaneously involved joints. Our motor control bandwidth is how fast we can move our own limbs or digits, which is much lower than the rate of motion we can perceive (Maclean 2008). Proprioceptive sensing occurs around 20–30 Hz, as compared to 10–10,000 Hz for tactile sensing. Control, however, saturates around 5–10 Hz, a limit determined by mechanical resonance—whether it is moving our eyes, fingers, or legs. Hasser and Cutkosky provide an example of how this has been modeled (Hasser and Cutkosky 2002). More detail on frequency ranges can be found in (Shimoga 1992).

Many psychophysical studies have sought to determine human capabilities such as the resolution with which we can track forces, i.e., maintain a desired force level with or without some form of feedback to us on the force we are providing; or similarly, our ability to follow a specified position trajectory in space. It has been revealed that humans are reported to track small forces in ideal conditions (visual feedback supplied) with errors ranging from 2 to 3 % when gripping (Mai et al. 1985) to 15 % when pushing against a normal surface (Srinivasan and Chen 1993). Performance degrades without visual feedback or access to texture (Lederman and Klatzky 2004); that is, both tactile and proprioceptive senses are apparently involved.

Sensorimotor control guides our physical motion in coordination with our touch sense (Lederman and Klatzky 1997). There is a different balance of "position" and "force" control when we are exploring an environment (e.g., lightly touching a surface) versus manipulating an object. Furthermore, neural correlates of haptic perception and motor control have been studied, including identified mapping between stimulus and brain cortex regions. The famous cortical homunculus provides the mapping between organs and brain areas (Marieb and Hoehn 2007).

1.1.2 Classification of Haptic Interfaces

A haptic interface is a feedback device that generates sensation to the skin and muscles, including a sense of touch, weight, and rigidity (Salisbury and Srinivasan 1997). It can be classified into two main categories: force-feedback device versus

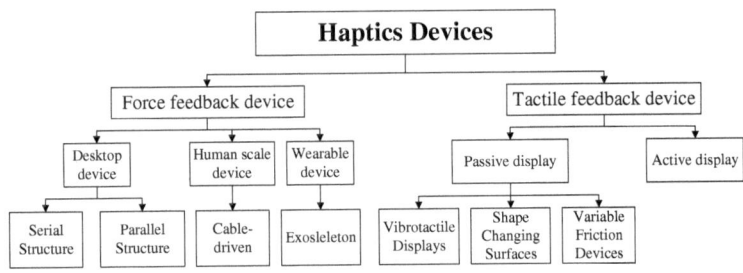

Fig. 1.2 A rough classification of haptic devices

tactile feedback device. Within each category, there can be further classification. Figure 1.1 shows one way of classifying haptic devices.

There are two types of force-feedback devices based on the direction of signal flows: impedance display and admittance display. The input and output of each type of device can be found in Hayward and MacLean (2007). The input and output of the corresponding haptic rendering can be designed, or it can be interconnected based on the virtual coupling concept (Colgate et al. 1995; Adams and Hannaford 1998).

There are many commercial haptic devices, including the Phantom series, Omega.X and Sigma.7, Virtuose, Haptic wand, and Haptic master. Many research prototypes have been developed. Please refer to (Hayward and MacLean 2007; Gosselin et al. 2008) for excellent surveys on the haptic devices.

Beside the types of devices in Fig. 1.2, there are also devices of hybrid types, which combine force-feedback devices with contact location display (Park et al. 2012) or skin-stretch devices (Guinan et al. 2013). In addition to direct contact devices, there are also some non-contact devices using ultrasonic wave (Hoshi et al. 2010), air pump device (Sodhi et al. 2013), or encountered-typed device (Yokokohji et al. 2004). Thermal display devices have also been studied and can be combined with tactile display devices (Ho and Jones 2007).

1.1.3 Overview of Haptic Interaction Systems

The origin of machine-mediated haptic interaction systems can be traced to one hundred years ago. From 1950s, the advent of computer and robotics technology produced a new research field, the haptic human–machine interaction (HHMI). The first prototype can be traced to the teleoperation system that facilitates manipulating an object in the area of nuclear engineering research and gets force feedback on the master side. The pioneer research started in the USA by Raymond Goertz and his colleagues at the Argonne National Laboratory (Goertz 1952) and in Europe by Vertut et al. (1976), and then becoming a rapidly expanding field. From then on, various haptic human–machine interaction systems have been developed to enable manipulation of virtual or remote environments through human haptic perception and feedback capabilities.

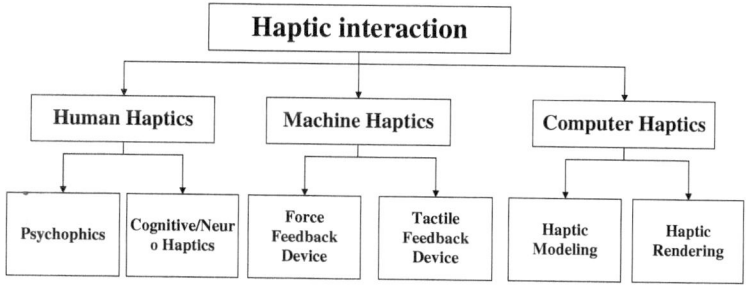

Fig. 1.3 Research branches of haptic human–machine interaction field

The goal of a HHMI system is to construct an interface between a human operator and a virtual or remote environment that mediated by a computer-generated avatar. Human operator, haptic interface or device, and virtual environment are three components of a haptic interaction system. Through haptic interaction systems, an human operator can feel the force-related properties from a virtual world or a remote physical world, including gravity force, inertia force, friction force, contact force, and reaction force. Along with graphic and auditory display, haptic-enabled multisensory display may provide a more powerful experience to approach the ultimate display (Sutherland et al. 1965), which can meet the three criteria of a high performance virtual reality system: imagination, immersion, and interaction.

A haptic interaction system is composed of two sub-systems: one real system (i.e., the user and the haptic device), and one virtual system (i.e., the tool and the environment). The tool acts as a virtual counterpart of the haptic device.

As shown in Fig. 1.3, research branches of HHMI field can be classified into three areas: human haptics, machine haptics, and computer haptics.

1.1.4 Example Applications

Various applications of HHMI systems have been developed, including medical applications, virtual prototyping, scientific visualization, tele-robotic manipulation (Brooks et al. 1990; Taylor et al. 1993), rehabilitation (Patton et al. 2004; Krebs et al. 2004; McLaughlin et al. 2005), and computer games (Lin and Otuday 2008) etc.

Harders summarized the roles and possible functions of haptics in medical applications (Harders 2008), including medical image data segmentation and visualization, tele-diagnosis, surgery and therapy planning, intra-operative support, rehabilitation, and medical education. For example, a number of surgical simulations systems have been developed for motor skill training, such as for laparoscopy, lumbar punctures, spine biopsy, dental simulation, and catheter insertion (Harders 2008).

Coquillart et al. summarized the roles and possible functions of haptics in virtual prototyping (Coquillart et al. 2008), which refers to the process of evaluating a new design of a system on a computer without the need of creating a physical prototype.

Virtual prototyping has been used for esthetical analysis, ergonomics evaluation, parts assembly, and manipulability study, in many industries, such as oil, automotive, and aeronautic industries. A large number of virtual prototyping tasks require force feedback, including human-in-the-loop assembling and disassembling, maintenance analysis, or ergonomic analysis (Lin and Otuday 2008).

Taylor summarized the roles and possible functions of haptics in scientific visualization (Taylor 2008). One famous example is the interactive molecular dynamics (IMD) system (Humphrey et al. 1996; Stone et al. 2001). This system permits manipulation of molecules in molecular dynamics simulations with real-time force feedback and graphical display. It enables scientists to pull on atoms in a running molecular-dynamic simulation and feel the force they are adding to the simulation. The particular strengths of haptic display have been twofold. First, haptic display is the only bidirectional channel between the scientist and computer. It enables the scientist to simultaneously sense the state of a system and control its parameters. Second, it has enabled the display of volumetric data sets without the problems of occlusion in visual displays. Recently, Paneels and Roberts provided a good survey on haptics applied to scientific visualization (Paneels and Roberts 2010).

Lin and Baxter provided a good summary of using haptics for modeling and creative processes (Lin and Baxter 2008). A creative process refers to any activity that involves translating imagination or conceptual design into more concrete forms. A system was designed to allow an artist to create 2.5D digital paintings, i.e., paintings with relief. It uses a physically based, deformable 3D brush model, which can be controlled by the artist via a haptic input device. Haptic interaction can considerably improve the ease of use and expressiveness of such systems.

Haptic technologies are still in search of a killer application, unlike computer graphics, which is widely used in movie and game industries. In recent years, adding force feedback to touch screen of mobile devices, such as mobile phones and tablet computers, in order to provide versatile tactile feelings has been an active topic of study. However, there are many technical challenges yet to be tackled, such as the size limit and power assumption.

1.2 Haptic Rendering: A Brief History

Salisbury et al. defined haptic rendering as the process of computing and generating forces in response to user interactions with virtual objects (Salisbury et al. 1995, 2004). Similar to graphic rendering, which aims to provide realistic visual display, haptic rendering aims to provide force display as realistically as possible.

Haptic rendering of the interaction between a virtual tool and a virtual environment consists of two tasks (Lin and Otuday 2008):

1. Compute and display the forces resulting from contact between the virtual tool and the virtual environment.
2. Compute the configuration of the virtual tool.

Fig. 1.4 Overview of a haptic rendering system

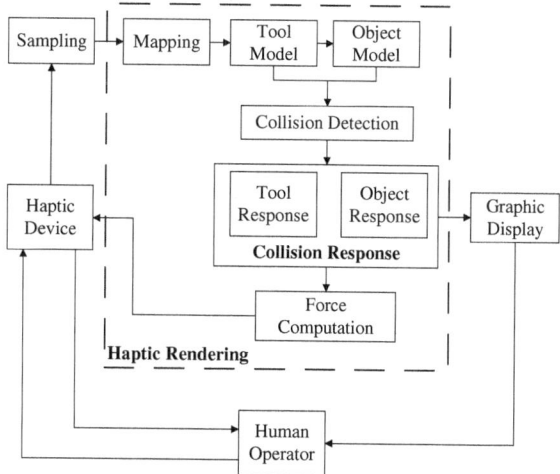

Figure 1.4 shows the key components of a haptic rendering system via an impedance-based display device, which include sampling the position of the haptic device and mapping it to the virtual tool, detecting collisions between the virtual tool and the virtual environment, computing collision responses (including contact force), and computing feedback force/torque for display.

An alternative is admittance-based rendering, where the haptic rendering system can be regarded as a programmable mechanical admittance that computes the desired device configuration, as the result of input forces (Lin and Otuday 2008). In that case, the device control should provide forces and implement a position control loop. As discussed by Adams and Hannaford (1998), impedance-based and admittance-based rendering systems present dual properties, and their designs can be addressed in a unified manner; therefore, in this book, we focus on impedance-based rendering.

Most earlier haptic rendering algorithms focus on 3-DoF haptic rendering. The haptic device and its avatar in the virtual environment is modeled as a point, i.e., the haptic interface point (HIP). The three-dimensional movement measured by the encoder of the haptic device was used to control the device avatar, which is defined as the surface contact point (SCP). Accordingly, only three-dimensional forces are feedback to the haptic device. Various methods have been proposed for 3-DoF haptic rendering, including the god-object method (Zilles and Salisbury 1995), the virtual proxy method (Ruspini et al. 1997), the intermediate method (Adachi et al. 1995; Mark et al. 1996). Basdogan and Srinivasan et al. extended the point tool to a line segment and proposed ray-based haptic rendering method (Basdogan et al. 1997; Ho et al. 2000). Ho et al. extended the object representation to triangle mesh (Ho et al. 1999; Gregory et al. 1999), implicit surfaces (Salisbury and Tarr 1997).

Since late 1990s, 6-DoF haptic rendering has been studied to address the multi-region contacts between a tool avatar, which can be of complex shape, and objects in a virtual environment (Fig. 1.5). Six-dimensional forces and torques exerted on the

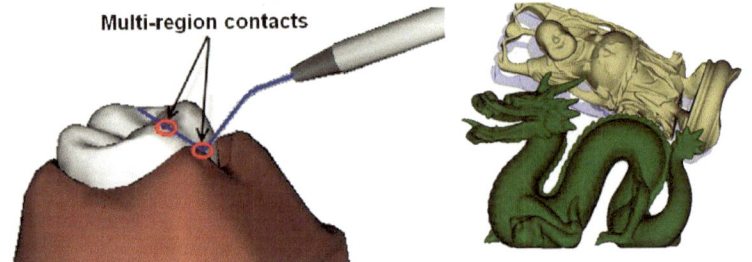

Fig. 1.5 Examples of multi-region contacts in 6-DoF haptic rendering

avatar are computed and feedback to the haptic device. McNeely et al. proposed the VPS (Voxmap and Point Shell) method to simulate rigid body contacts and applied the method to virtual maintenance of aircraft components (McNeely et al. 1999). Since then, many approaches have been introduced to simulate different scenarios including methods for rigid body contacts between polygonal models (Gregory et al. 2000; Otaduy and Lin 2008; Wang et al. 2013), and for deformable objects (Duriez et al. 2006; Luo and Xiao 2007; Barbič and James 2008; Wang et al. 2014), and topology changes (Courtecuisse et al. 2010). It is still a topic of active research to provide realistic simulation for 6-DoF haptic rendering (Otaduy et al. 2013).

1.3 6-DoF Haptic Rendering

Lin and Otuday proposed the general framework of 6-DoF haptic rendering as shown in Fig. 1.6 (Lin and Otuday 2008). The input is the six-dimensional configuration of the haptic device, and the output is the six-dimensional force and torque sent back to the haptic device.

The optimization problem is to find the proper collision response under the environment constraints that model the geometric and physical property of objects, and the force computation models the relationship between feedback force/torque and the simulated dynamic process. In this framework, the objective function for the optimization problem is not specified, nor the function for computing output

Fig. 1.6 General framework of 6-DoF haptic rendering (Lin and Otuday 2008)

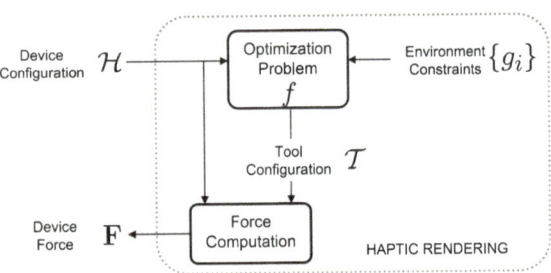

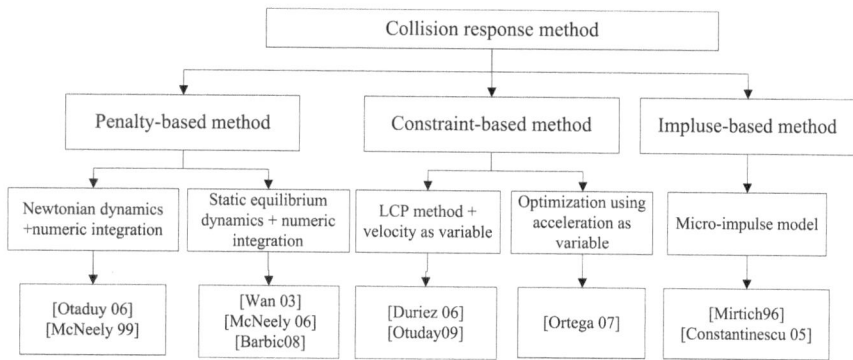

Fig. 1.7 Typical haptic rendering approaches classified by collision response methods

forces. The differences among various haptic rendering algorithms lie precisely in the design of these functions (Lin and Otuday 2008).

Contact constraints model the environment as algebraic equations in the configuration space of the tool, $g_i(T) \geq 0$. A configuration of the tool T_0 such that $g_i(T_0) = 0$ indicates that the tool is exactly in contact with the environment. Collision response exerts forces on the tool such that environment constraints are not violated.

Based on how collision response is handled, existing 6-DoF haptic rendering approaches can be classified into three kinds: penalty-based, constraint-based, and impulse-based approaches. Representative approaches are summarized and shown in Fig. 1.7.

There are other alternative ways to classify existing haptic rendering methods. For example, based on how objects are modeled, existing methods can be classified into method for triangle mesh, implicit surfaces, volume representation, spheres, or Nurbs surfaces (Lin and Otuday 2008).

1.3.1 Penalty-based Approach

In the penalty-based approach, contact constraints are modeled as springs whose elastic energy increases when the virtual tool penetrates into an object in the virtual environment. Penalty forces are computed as the negative gradient of the elastic energy, which pushes the graphic tool toward a non-penetrating configuration. Penalty forces can be defined in multiple ways, but the simplest is to consider a Hookean spring, where the contact penalty force is a linear function to the penetration depth between the tool and the objects, while the gradient of the function is the contact stiffness of the Hookean spring (Lin and Otuday 2008).

It is also common to apply penalty forces when objects come closer than a certain tolerance d. The addition of a tolerance has two major advantages: reducing object interpenetration and reducing the cost of collision detection, since distance computation is less costly than computation of penetration (Lin and Otuday 2008).

Penalty-based methods are attractive in several aspects: the force model is local to each contact and computationally simple, object inter-penetration is inherently allowed, and the cost of the numerical integration of computing the configuration of the virtual tool is almost insensitive to the complexity of contact configurations. This last property makes penalty-based methods well suited for interactive applications with high-frequency requirements, such as haptic rendering. In fact, penalty-based methods have been applied in many 6-DoF haptic rendering approaches (McNeely et al. 1999; Kim et al. 2003; Johnson and Willemsen 2003; McNeely et al. 2006; Otaduy and Lin 2006; Barbič and James 2007).

However, penalty-based methods also have some disadvantages. For example, there is no direct control over physical parameters, such as the coefficient of restitution, and friction forces are difficult to model. More importantly, geometric discontinuities in the location of contact points and/or normals lead to torque discontinuities (Lin and Otuday 2008).

Responsive penalty-based forces require the use of high stiffness values, which can compromise the stability of haptic display in direct rendering approaches and the stability of the simulation of the virtual environment in rendering approaches using virtual coupling. Implicit integration is known to provide high stability under larger combinations of mass and stiffness values, and it has been used for stable rigid body simulation with very stiff penalty forces, thus avoiding visually perceptible interpenetrations.

Many penalty-based methods utilized virtual coupling to maintain stability, and the main idea is to add an additional spring and damping between the virtual environment and the device (Colgate et al. 1995; Adams and Hannaford 1998), and thus to maintain the equivalent stiffness of the virtual environment is smaller than the maximum stiffness of the associated haptic device. The advantage of virtual coupling is to decouple the design of the VE and that of the haptic device; however, the fidelity of haptic rendering can be reduced because subtle changes of force signal will be filtered by virtual coupling when a tool slides along a surface with fine geometric details such as textured bumps and grooves.

There are two types of methods in the literature for computing the pose of the graphic tool: dynamic simulation with numerical integration and quasi-static equilibrium with integration (QSA approach). As pointed by Ortega et al., these methods may allow the virtual objects to interpenetrate, which may lead to missing collisions and the pop-through effect, where the graphic tool can traverse through thin objects or features, thereby degrading the perception of geometric details (Ortega et al. 2007).

When the allowed free space for the virtual tool is very small, small computation inaccuracy can make it penetrate into the surface of a cavity or even vibrate between two surfaces. Furthermore, a virtual coupling is often used to maintain the stability of the haptic feedback, which can lead to incorrect forces by modifying the force orientation applied to the user. As details of the feedback force are filtered by virtual coupling, the human operator cannot feel the subtle force changes when the tool interacts with fine features of a complex object.

1.3.2 Constraint-based Approach

Another way of modeling environment contact constraints is the constraint-based approach using the Lagrange multipliers. Constraint-based methods attempt to constrain the pose of the displayed virtual tool, called *graphic tool*, to be free of penetration, while the actual virtual tool, called *haptic tool*, can penetrate into objects. In contrast to the penalty-based approach, constraint-based approach is an analytical and global method for computing the collision response.

Constraint-based methods with Lagrange multipliers handle all concurrent contacts as a single optimization problem and attempt to find contact forces that produce physically and geometrically valid motions. As opposed to penalty-based methods, constraint-based methods allow, for example, for relatively easy inclusion of accurate friction models. The method of Lagrange multipliers allows for an exact enforcement of contact constraints $g_i(T) \geq 0$ by modeling workless constraint forces $\mathbf{F}_C = \mathbf{J}^T \lambda$ normal to the constraints. Constraint forces are added to regular forces of the dynamic equations of a colliding object (e.g., the tool). Then, constraints and dynamics are formulated in a joint differential-algebraic system of equations. The "amount" of constraint force λ is the unknown of the system, and it is solved such that constraints are enforced.

Typically, contact constraints are non-linear, but solving constrained dynamics systems can be accelerated by linearizing the constraints. The addition of constraints for non-sticking forces $\lambda \geq 0$, $\lambda T_g(\mathbf{q}) = 0$ yields a linear complementarity problem (LCP) (Cottle et al. 1992).

Early constraint-based techniques pause the simulation at all collision events and formulate a linear complementarity problem to solve for collision forces and object accelerations. Some researchers have integrated this approach into haptic rendering, but they have tested it only on relatively simple benchmarks, due to the typically high cost of variable time-stepping. Others have developed constraint-based techniques with fixed time-stepping, but they may suffer from drift since the constraints are expressed in velocities.

There are other variants of the problem, for example, by allowing sticking forces through equality constraints, or differentiating the constraints and expressing them on velocities or accelerations.

Constraint-based methods are computationally expensive, even for the linearized system, and the solution of constrained dynamics and the definition of constraints (i.e., collision detection) are highly intertwined.

Constraint-based methods achieve accurate simulation by modeling the normal and friction contact constraints as a linear complementary problem, which is time consuming to solve due to iteratively solving differential-algebraic equations or carrying out mesh-based continuous collision detection (CCD). A multi-rate architecture is introduced to solve this problem, i.e., the set of non-penetration constraints are computed and updated at a relatively slow thread, and adaptation or interpolation of contact configurations is carried out to achieve a 1-kHz haptic rate in another thread. As Ortega et al. (Ortega et al. 2007) pointed out, "this might lead to missing some high frequency details when the user slides rapidly on the surface of the environment obstacles."

1.3.3 Impulse-based Approach

Impulse-based techniques pause haptic the simulation at all collision events and resolve contacts based solely on impulses (Mirtich 1996). The major drawback is that resting contact is modeled by multiple micro-collision events, which is inaccurate. Chang and Colgate integrated passive impulse-based techniques with haptic rendering and stressed the need for other methods to handle resting contact (Chang and Colgate 1997). Constantinescu et al. proposed the combination of penalty forces with impulsive response (Constantinescu et al. 2005). They have proven the passivity of multiple impulses applied simultaneously using Newton's restitution law.

1.4 Fine Manipulation and Its Computational Challenges

In this section, we characterize fine manipulation and classify different types of fine manipulation based by several examples. We also describe two main features of fine manipulation and its computational challenges for haptic rendering.

1.4.1 Characterization of Fine Manipulation

Fine manipulations refer to manipulations that involve small movement and/or accurate force control of a tool interacting with objects. It is pervasive—there are many examples of fine manipulation in industry, such as assembly and suturing, and in everyday life, such as grasping an egg, eating food with chopsticks or fork and knife, and playing a violin. In surgical operations, accurate control of force or movement of a surgical instrument, such as a needle, is necessary to ensure precision and minimize unintended damage.

According to the touch media, fine manipulation can be roughly classified into two types: hand-mediated interaction and tool-mediated interaction. In the former type, a human hand directly touches and manipulates an object, as shown by the examples in Fig. 1.8; while in the latter case, a hand manipulates an object through a tool, as shown by the examples in Fig. 1.9. In this book, we focus on tool-mediated fine manipulation.

Figure 1.10 illustrates the components of fine manipulation, which include two aspects: human aspect and task aspect. A human operator needs fine motor skill and hand-eye coordination skill to accomplish a fine manipulation task. Fine motor skill

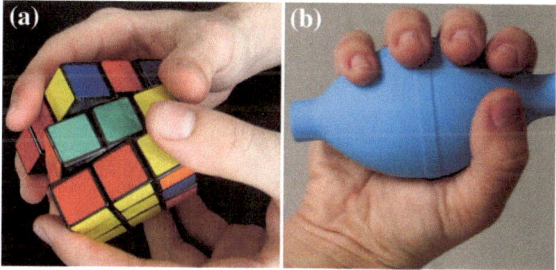

Fig. 1.8 Examples of hand-mediated fine manipulations. **a** Solving a magic cube. **b** Deforming a balloon

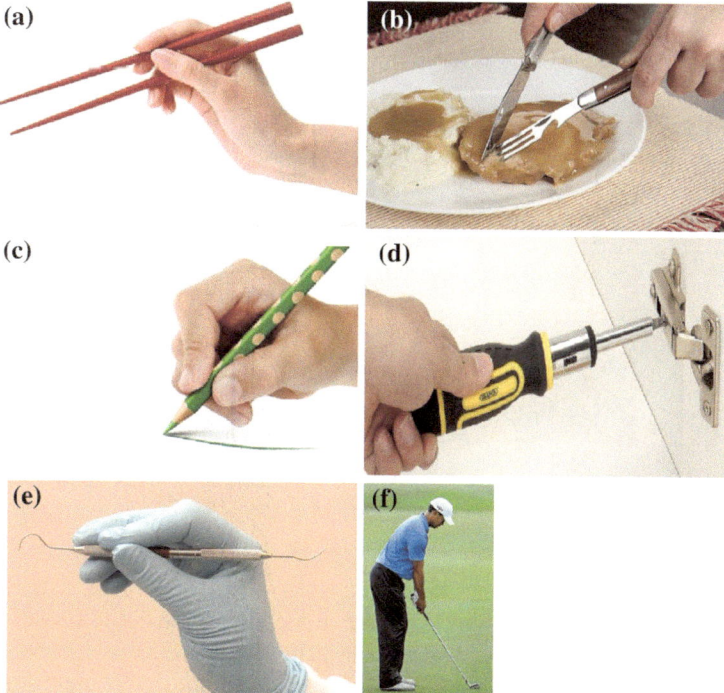

Fig. 1.9 Examples of tool-mediated fine manipulations. **a** Using chopsticks to pick a peanut. **b** Coordinating a knife and a fork to cut beef tendons. **c** Grasping a pen to draw a curve. **d** Using a screwdriver to do assembly. **e** Manipulating a dental instrument. **f** Controlling of a golf-club

refers to using joints and muscles such as fingers and wrist to produce precise, independent, and simultaneous motion and force output to conduct the manipulation.

From the task aspect, a fine manipulation task is often characterized by narrow operation space and the need to use haptic feedback of fine-feature interaction

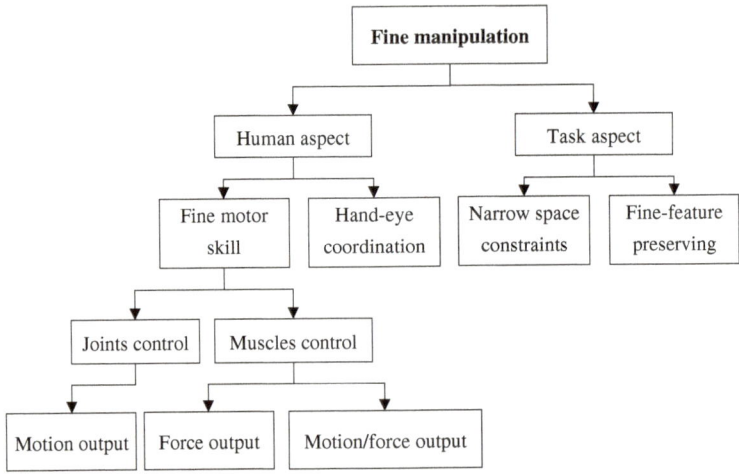

Fig. 1.10 Components of fine manipulation

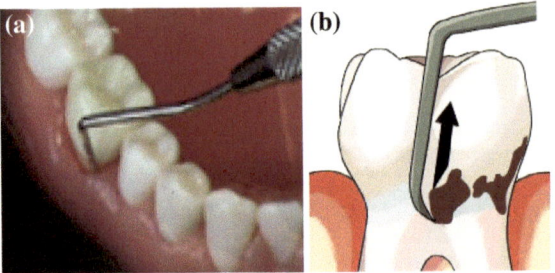

Fig. 1.11 Two characteristics in periodontal operation: **a** narrow cavities produced by teeth and gingival; **b** small-sized calculi located within the pocket

between the tool and the objects. For example, as shown in Fig. 1.11, in periodontal operation, the dental probe must goes into a periodontal pocket to detect whether any calculi are located in the pocket. The periodontal pocket is a narrow cavity between the tooth and the surrounding gingiva. Furthermore, there are many small-sized calculi located within the pocket; a dentist needs to detect them and remove them based on haptic feedback. Similarly, for the assembly of an aircraft engine shaft in Fig. 1.12, a splined shaft needs to be inserted into a splined hole, and the force feeling against fine features is used to ensure the success of insertion.

Fine manipulation presents a major challenge for haptic rendering: how to satisfy narrow space constraints on the manipulated tool and fine-feature preserving requirement in handling complex contact scenarios to achieve high accuracy as well as stability in haptic rendering, all in 1 kHz.

Fig. 1.12 Two characteristics in aircraft engine assembly: **a** narrow cavities produced by the small clearance between a spline shaft and a spline hole; **b** small-sized features of the spline teeth

1.4.2 Narrow Space Constraints

Narrow space constraints are due to the fact that in many delicate manipulation tasks, the space for the tool to maneuver is very small (Wang et al. 2013). For example, in dental probing, the probe (tool) is constrained by both the tooth and the gingival in a narrow cavity between them. As shown in Fig. 1.13, frequent constraint changes or contact switches occur during the tool's movement. As a result, in haptic rendering, a small movement (translation and/or rotation) of the haptic tool will lead to a change of contact constraints, in terms of a change of the number of contact regions or of the contact constraint type within each region (Fig. 1.14). The frequent constraint changes or contact switches pose a great challenge for most haptic rendering methods.

Narrow space is a relative concept, as illustrated by the tool-in-hole example in Fig. 1.15. We define a rough criterion for narrow space in haptic rendering based on the following relation: the maximum clearance δ_c between the graphic tool and the object surfaces bounding the space is smaller than the maximum displacement δ_h of the haptic tool per time step. It should be noted that for a tool with long length, small rotation angle about the center may lead to large displacement on the tip. This can lead to frequent change of contact scenarios under small rotations.

6-DoF haptic rendering of fine manipulation in narrow space has a lot of potential applications. As shown in Fig. 1.16, there are many other simulation scenarios that narrow space caused by cluttered objects or tight constraints.

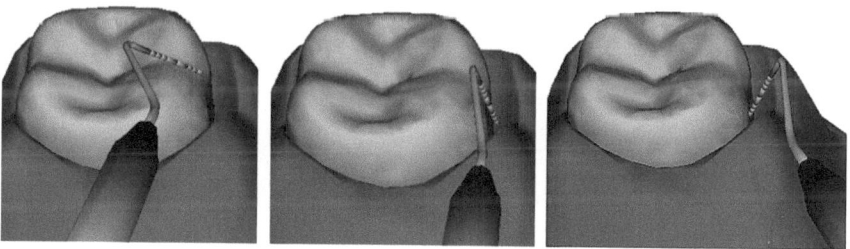

Fig. 1.13 Narrow space between a tooth and the gingiva leads to frequent contact switches

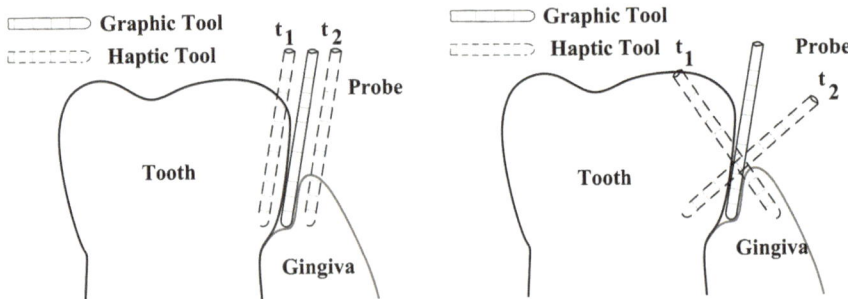

Fig. 1.14 Small movement of the tool leads to a change of contact constraint status and force/torque discontinuities. Copyright © IEEE. All rights reserved. Reprinted, with permission, from Wang et al. (2013)

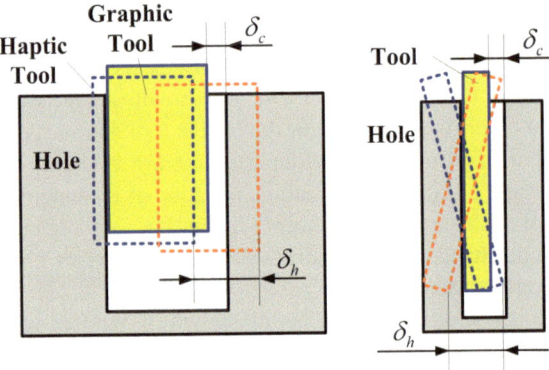

Fig. 1.15 Illustration of the narrow space criterion by using a tool-in-hole example (The *yellow* and *solid* tool refers to the graphic tool, and the *dashed* tools refer to the haptic tool in two adjacent time steps)

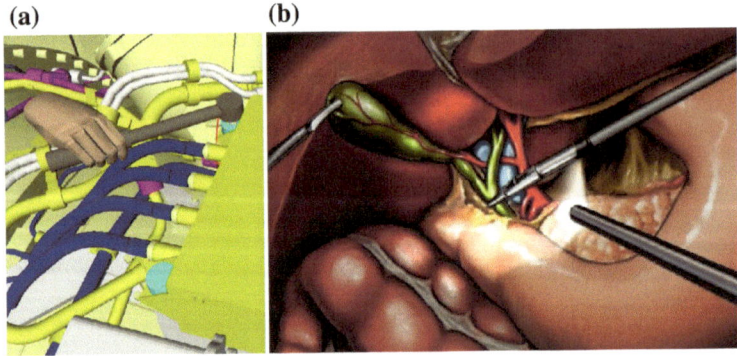

Fig. 1.16 Example simulation scenarios involving manipulation in narrow space. **a** Mechanical structures assembly. **b** Laparoscopic operation

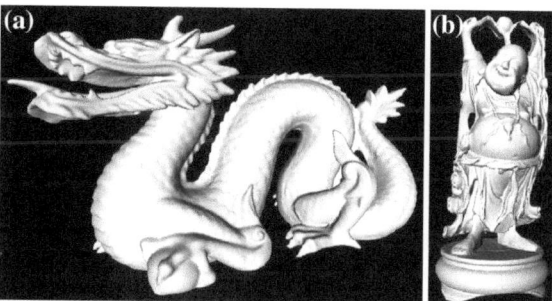

Fig. 1.17 Example objects involving fine features. **a** Small features on the surface of a dragon. **b** Small features on the surface of the Happy Buddha model (https://graphics.stanford.edu/courses/cs348b-competition/cs348b-03/buddha/)

1.4.3 Simulation of Fine Features

Fine features refer to small geometric features on an object. As shown in Fig. 1.17, when a human operator uses a tool to slide over objects with fine features, the subtle force feedback provides important cues to the operator about the physical interaction. In haptic rendering, it is important to model and simulate contact constraints between the virtual tool and such fine features on virtual objects in fine details in order to achieve realistic force simulation.

A fine feature can be characterized in two aspects. First is the ratio of the feature's absolute size to the movement range of a haptic device per time step. As the size of the features is small, small movement of the haptic device across the feature will lead to change of force torque signals. This kind of fine features imposes stringent requirements for the haptic rendering method.

In the second aspect, objects with special shapes can be treated as fine features, such as thin object and sharp features. For example, for the thin object simulation such as removing a calculus as shown in Fig. 1.11b, the thickness of the calculus is very small, typical collision detection method will produce tunneling effect. To achieve a realistic haptic simulation, pop-through (i.e., tunneling effect) against thin-objects or small-size features on the object's surface should be avoided.

In this book, we introduce a systematic approach for 6-DoF haptic simulation of fine manipulation that addresses the challenges described above.

References

Adachi Y, Kumano T, Ogino K (1995) Intermediate representation for stiff virtual objects. In Proceedings of virtual reality annual international symposium, pp 203–210

Adams RJ, Hannaford B (1998) A two-port framework for the design of unconditionally stable haptic interfaces. In Proceedings IEEE/RSJ international conference intelligence robots system, pp 1254–1259

Barbič J, James DL (2007) Time-critical distributed contact for 6-DoF haptic rendering of adaptively sampled reduced deformable models. In: Proceedings of ACM SIGGRAPH/ Eurographics symposium on computer animation. Eurographics Association, Aire-la-Ville, Switzerland, pp 171–180

Barbič J, James DL (2008) Six-DoF haptic rendering of contact between geometrically complex reduced deformable models. IEEE Trans Haptics 1(1):39–52

Basdogan C, Ho C, Srinivasan MA (1997) A ray-based haptic rendering technique for displaying shape and texture of 3D objects in virtual environments. Winter Annu Meeting of ASME 61:77–84

Brooks Jr FP, Ouh-Young M, Batter JJ, Kilpatrick PJ, Baskett F (ed) (August 1990) Project GROPE—haptic displays for scientific visualization. In Proceedings computer graphics (SIGGRAPH), vol. 24, pp 177–185

Chang B, Colgate JE (1997) Real-time impulse-based simulation of rigid body systems for haptic display. In: Proceedings ASME dynamic system control division, pp 145–152

Colgate JE, Stanley MC, Brown JM (1995) Issues in the haptic display of tool use. In: Proceedings of IEEE/RSJ international conference of intelligence robots system, pp 140–145

Constantinescu D, Salcudean SE, Croft EA (2005) Haptic rendering of rigid contacts using impulsive and penalty forces. IEEE Trans Robot 21(3):309–323

Coquillart S, Ortega M, Tarrin N (2008) Chapter 22 Virtual prototyping, In: Lin MC, Otaduy M (eds) Haptic rendering: foundations, algorithms, and applications, AK Peters Ltd, Massachusetts 2008

Cottle RW, Pang JS, Stone RE (1992) The linear complementarity problem. Academic Press, San Diego

Courtecuisse H, Jung H, Allard J, Duriez C, Lee DY, Cotin S (2010) GPU-based real-time soft tissue deformation with cutting and haptic feedback. Prog Biophys Mol Biol 103:159–168

Duriez C, Dubois F, Kheddar A, Andriot C (2006) Realistic haptic rendering of interacting deformable objects in virtual environments. IEEE Trans Vis Comput Graphics, pp 36–47

Florian G, Claude A, Joan S, Javier M (2008) Large workspace haptic devices for human-scale interaction: a survey. EuroHaptics 2008, Madrid, Spain, 10–13 June, 2008

Goertz RC (1952) Fundamentals of general-purpose remote manipulators. Nucleonics 10:36–42

Gregory A, Lin M, Gottschalk S, Taylor R (1999) H-COLLIDE: a framework for fast and accurate collision detection for haptic interaction. In: Proceedings of virtual reality conference 1999. IEEE Computer Society, Washington, DC, pp 38–45

Gregory A, Mascarenhas A, Ehmann S, Lin MC, Manocha D (2000) 6-DOF haptic display of polygonal models. In: Proceedings of IEEE visualization conference 2000, pp 139–146

Guinan AL, Hornbaker NC, Montandon AJ, Doxon, Provancher WR (2013) Back-to-back skin stretch feedback for communicating five degree-of-freedom direction cues, In: Proceeding of world haptics conference (WHC)

Harders M (2008) Chapter 24, Haptics in medical applications. In: Lin MC, Otaduy M (eds) Haptic rendering: foundations, algorithms, and applications. A K Peters Ltd, Massachusetts

Hasser CJ, Cutkosky MR (2002) System identification of the human hand grasping a haptic knob. In: Proceedings of the 10th symposium on haptic interfaces for virtual environment and teleoperator systems. IEEE, Los Alamitos, CA, pp 171–180

Hayward V, MacLean KE (2007) Do it yourself haptics—Part I. IEEE Robotics and Autom Soc Mag 14(4):88–104

Ho C, Basdogan C, Srinivasan MA (1999) An efficient haptic rendering technique for displaying 3D polyhedral objects and their surface details in virtual environments. Presence Teleoperators Virtual Environ 8(5):477–491

Ho C, Basdogan C, Srinivasan MA (2000) Ray-based haptic rendering: interactions between a line probe and 3D objects in virtual environments. Int J Robot Res 19(7):668–683

Ho HN, Jones LA (2007) Development and evaluation of a thermal display for material identification and discrimination. ACM Trans Appl Percept 4:1–24

Hoshi T, Takahashi M, Iwamoto T, Shinoda H (2010) Noncontact tactile display based on radiation pressure of airborne ultrasound. IEEE Trans Haptics, 3:155

Humphrey W, Dalke A, Schulten K (1996) VMD: visual molecular dynamics. J Mol Graph 14:33–38

Johnson DE, Willemsen P (2003) Six degree of freedom haptic rendering of complex polygonal models. In: Proceedings of haptics symposium. IEEE Computer Society, Washington, DC, pp 229–235

Jones LA (1998) Perception and control of finger forces. In: Proceedings haptics symposium, ASME dynamic systems and control division DSC-64, pp 133–137

Jones Lynette, Tan Hong Z (2013) Application of psychophysical techniques to haptic research. IEEE Trans Haptics 6(3):268–284

Kim YJ, Otaduy MA, Lin MC, Manocha D (2003) Six-degree-of-freedom haptic rendering using incremental and localized computations. Presence 12(3):277–295

Klatzky RL, Lederman SJ (1999) Tactile roughness perception with a rigid link inter posed between skin and surface. Perception Psychophys 61(4):591–607

Klatzky RL, Lederman SJ (2003) Touch. In: Weiner IB (ed) Handbook of psychology: experimental psychology. Wiley, New York, pp 147–176

Krebs H, Ferraro M, Buerger S, Newbery M, Makiyama A, Sand-mann M, Lynch D, Volpe B, Hogan N (2004) Rehabilitation robotics: pilot trial of a spatial extension for MIT-manus. J Neuro Eng Rehabil 1:5

Lederman S, Klatzky R (1997) Designing haptic interfaces for teleoperational and virtual environments: should spatially distributed forces be displayed to the fingertip? In: Proceedings of the ASME dynamic systems and control division. ASME, New York

Lederman SJ, Klatzky RL (2004) Multisensory texture perception. In: Stein BE, Spence C, Calvert G (eds) The handbook of multisensory processes. MIT Press, Cambridge, pp 107–122

Lin M, Baxter W (2008) Chapter 26, Modeling and creative processes. In: Lin MC, Otaduy M (eds) Haptic rendering: foundations, algorithms, and applications, A K Peters Ltd, Massachusetts

Lin MC, Otaduy M (2008) Haptic rendering: foundations, algorithms, and applications. A K Peters Ltd, Massachusetts

Luo Q, Xiao J (2007) Contact and deformation modeling for interactive environments. IEEE Trans Rob 23(3):416–430

MacLean KE (2008) Haptic interaction design for everyday interfaces. Rev Hum Factors Ergon 4:149–194

Mai N, Avarello M, Bolsinger P (1985) Maintenance of low isometric forces during prehensile grasping. Neuropsychologia 23:805–812

Marieb E, Hoehn K (2007) Human anatomy and physiology, 7th edn. Pearson Benjamin Cummings, San Francisco

Mark WR, Randolph SC, Finch M, Verth JMV, Taylor II RM (1996) Adding force feedback to graphics systems: issues and solutions. In: Holly R (ed) Proceedings of SIGGRAPH, computer graphics proceedings, annual conference series, Addison Wesley, Reading, MA, pp 447–452

McLaughlin ML, Rizzo AA, Jung Y, Peng W, Yeh S, Zhu W (2005) Haptics-enhanced virtual environments for stroke rehabilitation. In Proceedings of IPSI, 2005

McNeely W, Puterbaugh K, Troy J (1999) Six degree-of-freedom haptic rendering using voxel sampling. In: Alyn R (ed) Proceedings of SIGGRAPH'99, computer graphics proceedings, annual conference series, Addison Wesley Longman, Reading, MA, pp 401–408

McNeely W, Puterbaugh K, Troy J (2006) Voxel-based 6-DoF haptic rendering improvements. Haptics-e 3(7)

Mirtich B (1996) Impulse-based dynamic simulation of rigid body systems. PhD thesis, University of California, Berkeley

Ortega M, Redon S, Coquillart S (2007) A six degree-of-freedom god-object method for haptic display of rigid bodies with surface properties. IEEE Trans Vis Comput Graphics, 13 (3):458–469

Otaduy MA, Garre C, Lin MC (2013) Representations and algorithms for force-feedback display. Proc IEEE 101(9):2068–2080

Otaduy MA, Lin MC (2006) A modular haptic rendering algorithm for stable and transparent 6-DoF manipulation. IEEE Trans Robot 22(4):751–762

Paneels S, Roberts JC, (2010) Review of designs for haptic data visualization. IEEE Trans Haptics 3(2):119–137

Park J, Doxon AJ, Provancher WR, Johnson DE, Tan HZ (2012) Haptic edge sharpness perception with a contact location display. IEEE Trans Haptics 5(4):323–331

Patton J, Dawe G, Scharver C, Mussa-Ivaldi F, Kenyon R (2004) Robotics and virtual reality: the development of a life-dized 3-D system for the rehabilitation of motor function. In: Proceedings of IEEE engineering medical biology society, IEEE Computer Society, Washington D.C, pp 4840–4843

Ruspini DC, Kolarov K, Khatib O (1997) The haptic display of complex graphical environments. In: Turner W (ed) Proceedings of SIGGRAPH 97, computer graphics proceedings, annual conference series, Addison Wesley, Reading, MA, pp 345–352

Salisbury K, Brock D, Massie T, Swarup N, Zilles C (1995) Haptic rendering: programming touch interaction with virtual objects. In: Proceedings of the symposium on interactive 3D graphics, pp 123–130. New York: ACM Press

Salisbury JK, Srinivasan MA (1997) Phantom-based haptic interaction with virtual objects. IEEE Comput. Graphics Appl 17(5):6–10

Salisbury JK, Tarr C (1997) Haptic rendering of surfaces defined by implicit functions. Proc ASME 61:61–67

Salisbury K, Barbagli F, Conti F (2004) Haptic rendering: introductory concepts. IEEE Comput Graphics Applications Mag 24(2):24–32

Shimoga KB (1992) Finger force and touch feedback issues in dexterous telemanipulation. In: Proceedings of fourth annual conference on intelligent robotic systems for space exploration. IEEE Computer Society, Los Alamitos, CA, pp 159–178

Sodhi R, Poupyrev I, Glisson M, Israr A (2013) AIREAL: interactive tactile experiences in free air. In: Proceedings of ACM SIGGRAPH 2013

Srinivasan MA, Chen JS (1993) Human performance in controlling normal forces of contact with rigid objects. In: Advances in Robotics, Mechatronics, and Haptic Interfaces, vol 49. ASME

Stone JH, Gullingsrud H, Schulten K (2001) A system for interac-tive molecular dynamics simulation. In: Symposium on interactive 3D graphics, ACM Press, New York, pp 191–194

Sutherland IE (1965) The ultimate display. Proceedings of IFIP congress, pp 506–508

Tan HZ, Adelstein BD, Traylor R, Kocsis M, Hirleman ED (2006) Discrimination of real and virtual high-definition textured surfaces. In: Proceedings of symposium on haptic inter faces for virtual environment and teleoperator systems (HAPTICS'06), IEEE Computer Society, Los Alamitos, CA, p 1

Taylor RM, Robinett W, Chi VL, Brooks Jr FP, Wright WV, Williams RS, Snyder EJ (1993) The nanomanipulator: a virtual-reality in-terface for a scanning tunneling microscope. In: James TK (ed) Proceedings of SIGGRAPH 93, computer graphics proceedings, annual conference series, ACM Press, New York, pp 127–134

Taylor R (2008) Chap. 23, Haptics for scientific visualization, In: Lin MC, Otaduy M, Haptic rendering: foundations, algorithms, and applications, A K Peters Ltd, Massachusetts

Vertut J, Marchal P, Debrie G et al (1976) MA-23 bilateral servomanipulator system, Trans Am Nucl Soc, vol. 24

Wang D, Zhang X, Zhang Y, Xiao J (2013) Configuration-based optimization for six degree-of-freedom haptic rendering for fine manipulation. IEEE Trans Haptics 6(2):167–180

Wang D, Shi Y, Liu S, Zhang Y, Xiao J (2014) Haptic simulation of organ deformation and hybrid contacts in dental operations. IEEE Trans Haptics 7(1):48–60

Yokokohji Y, Muramori N, Sato Y, Yoshikawa T (2004) Designing an encountered-type haptic display for multiple fingertip contacts based on the observation of human grasping behavior, In: Proceedings of haptic interfaces for virtual environment and teleoperator systems, HAPTICS'04. 27–28 March

Zilles CB, Salisbury JK (1995) A constraint-based god-object method for haptic display. In: Proceedings IEEE/RSJ intelligence conference intelligent robots and systems, August 1995

Chapter 2
Configuration-based Optimization Approach

To tackle computational challenges in 6-DoF haptic rendering of fine manipulation, in this book we present a novel constraint-based approach: the configuration-based optimization approach. In this chapter, we introduce the basics of the configuration-based optimization approach to 6 DoF haptic rendering of rigid body in contact. First, we provide an overview of the approach in Sect. 2.1, and then we introduce each component in the computational pipeline in Sects. 2.2–2.4, through which we explain the differences between this approach and other haptic rendering approaches. In Section 2.5, we introduce experimental results that validate the approach.

2.1 Overview of the Approach

We first introduce the problem and the motivation of the approach. Next, we outline the framework of the approach with the basic components in a flowchart. We conclude this section by summarizing the main contributions of the approach.

2.1.1 Problem Formulation

We introduce the following terms to facilitate description:

- *Haptic tool*: It is the avatar in a virtual environment of joystick handle of the haptic device held by a user in the physical world; it reflects the exact location of the joystick handle as the user moves the handle and can penetrate the surface of a rigid object in the virtual world. In 3-DoF haptic rendering, this tool has been named as the haptic interface point (HIP) (Ho et al. 1999), or haptic handle (Lin and Otaduy 2008).

Electronic supplementary material The online version of this article (doi:10.1007/978-3-662-44949-3_2) contains supplementary material, which is available to authorized users.

- *Graphic tool*: It is the graphic display of the haptic tool that forms a contact with an object but does not penetrate the object to simulate the physical reality of contact. This term is also called the god-object (Zilles and Salisbury 1995), virtual proxy (Ruspini et al. 1997), surface contact point (SCP) (Ho et al. 1999), or virtual tool (Ortega et al. 2007).
- *Collision response*: The process is to compute the pose of the graphic tool in contact and to simulate the contact force and torque (Lin and Otaduy 2008).

The motivation of the configuration-based optimization approach is to tackle the problem of computing the pose of the graphic tool when the haptic tool penetrates into a virtual object so that its pose cannot be directly used as the pose of the graphic tool, as the latter simulates contact with the virtual object. Intuitively, the relation between the haptic tool and the corresponding graphic tool can be illustrated by the magnet metaphor shown in Fig. 2.1, where the haptic tool inside the virtual object can be viewed as a magnet, the object's surface can be viewed as a metal container, and the graphic tool can be viewed as attracted by and following the haptic tool as close as possible but cannot enter the container (i.e., cannot penetrate through the object). If the haptic tool is outside the virtual object and in the free space, the graphic tool collocates with the haptic tool.

The configuration-based optimization approach takes as input the configuration of the haptic tool and computes the corresponding configuration of the graphic tool as the result of optimizing the configuration of the graphic tool directly rather than optimizing the acceleration or velocity (Duriez et al. 2006; Ortega et al. 2007) to avoid inaccuracies in the configuration of the graphic tool due to numerical integration.

In addition, the configuration-based optimization approach allows us to compute the tool configuration without incorporating typical mesh to mesh continuous collision detection [such as FAST or C2A (Redon 2004; Zhang et al. 2006; Tang et al. 2009)] and thus avoid the associated computational burden. For example, in (Ortega et al. 2007), the final configuration of the graphic tool is computed by continuous collision detection (CCD) instead of optimization. In each simulation

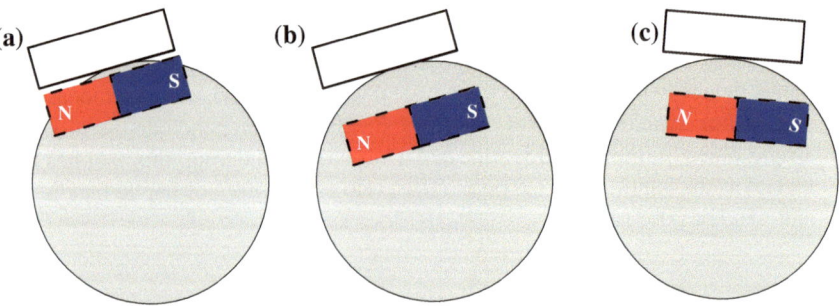

Fig. 2.1 The magnet metaphor. **a** Haptic tool entering the virtual object. **b** Graphic tool following the translating haptic tool. **c** Graphic tool following the rotating haptic tool

cycle, an acceleration-based optimization and a subsequent integration are executed before the CCD, because the result of integration cannot maintain non-penetration between the moving graphic tool and the other objects. However, the configuration-based optimization approach can provide a solution directly satisfying the non-penetration constraints without requiring a post-optimization CCD module.

For a tool and objects modeled in polygonal meshes, the configuration-based optimization approach produces high-fidelity rendering (Wang et al. 2011), but efficient operations are limited to small-region contacts and a tool of simple shape. In order to treat more complex contact scenarios efficiently, we consider the tool and objects modeled in sphere trees. Sphere trees facilitate a uniform expression of constraints for complex contacts involving both convex and concave geometric features, which simplifies configuration-based optimization. Detailed analysis about the reason for adopting sphere trees for constraint modeling will be explained in Sect. 2.2.1.

Since we focus on haptic simulation of multi-region contacts with frequent contact switches, we do not assume very large and fast movements of the haptic tool when the haptic tool intersects with an object, that is, we assume quasi-static movements. This assumption is reasonable for related applications, such as simulating dental operations, where a dentist moves the dental tool carefully and slowly in a patient's mouth.

2.1.2 Framework

The input and output of the optimization problem are:

- Given: (1) the 6D configuration of the haptic tool at the previous and current time step $\left(\mathbf{q}_h^{t-1}, \mathbf{q}_h^t\right)$, and (2) the 6D configuration of the object (e.g., a tooth), the geometric model and the physical property (e.g., stiffness and friction coefficient) of the object.
- Compute: (1) The 6D configuration of the graphic tool at the current time step $\left(\mathbf{q}_g^t\right)$ while ensuring no penetration between the graphic tool and the object, (2) the 6D feedback force and torque $\mathbf{F}^t$.

We use the following configuration-based optimization model (As shown in Fig. 2.2)

$$\begin{cases} \text{Min:} & f(\mathbf{q}_g^t, \mathbf{q}_h^t) \\ \text{Subject to:} & V_O \cap V_T = \varnothing \end{cases} \tag{2.1}$$

where

$$\begin{cases} \mathbf{q}_h = \left[x_h, y_h, z_h, \gamma_h, \beta_h, \alpha_h\right]^{\mathrm{T}} \\ \mathbf{q}_g = \left[x_g, y_g, z_g, \gamma_g, \beta_g, \alpha_g\right]^{\mathrm{T}} \end{cases} \tag{2.2}$$

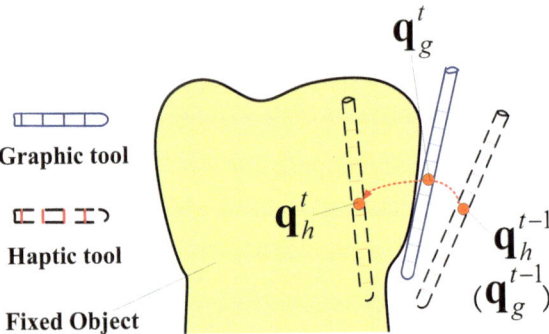

In Eq. (2.1), V_O denotes the volume of the object and V_T denotes the volume of the graphic tool. The constraint is to maintain exact contact between the graphic tool and the object, i.e., there is no perceptible penetration and no separation. The key challenge in this optimization model is how to formulate the constraint in Eq. (2.1), which will be discussed in Sect. 2.2.

In Eq. (2.1), the objective function $f(\bullet)$ is a distance metric to describe the difference between configurations of the haptic tool and the graphic tool. In related literature (Zhang 2009), different kinds of distance metrics were introduced, but the time cost for computing each of those metrics is far greater than 1 ms, which does not satisfy the update rate of 1 kHz, as required by haptic rendering, called *haptic rate*.

In order to make sure that the graphic tool follows the movement of the haptic tool without unnatural movement as the haptic tool translates and rotates, we imagine that a six-dimensional spring connects the graphic tool to the haptic tool and define the following deformation energy of the spring as the objective function to minimize:

$$f(q_g^t) = \sum_{j=1}^{3} \frac{1}{2} k_t (q_{gj}^t - q_{hj}^t)^2 + \sum_{j=4}^{6} \frac{1}{2} k_r (q_{gj}^t - q_{hj}^t)^2 \qquad (2.3)$$

where k_t and k_r are translational and rotational stiffness, respectively. For implementation of the proposed model, these two values are chosen as $k_t = 1$ N/mm and $k_r = 1,000$ mNm/rad to fully exploit the ability of the haptic device used.

The above objective function can be further represented in the following matrix form:

$$f(\mathbf{q}_g^t, \mathbf{q}_h^t) = \frac{1}{2} (\mathbf{q}_g^t - \mathbf{q}_h^t)^T \mathbf{G} (\mathbf{q}_g^t - \mathbf{q}_h^t) \qquad (2.4)$$

where $\mathbf{G}$ is a diagonal 6×6 stiffness matrix. $\mathbf{q}_h^t = (x_h^t, y_h^t, z_h^t, \gamma_h^t, \beta_h^t, \alpha_h^t)$, and $\mathbf{q}_g^t = (x_g^t, y_g^t, z_g^t, \gamma_g^t, \beta_g^t, \alpha_g^t)$.

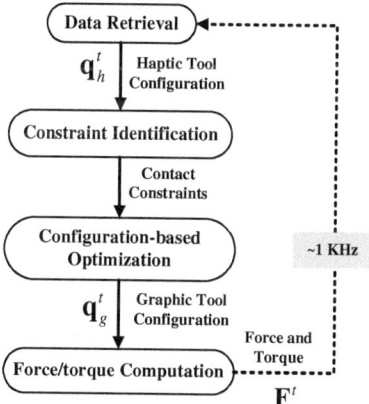

Fig. 2.3 Flowchart of the configuration-based optimization approach to 6-DoF haptic rendering. Copyright © IEEE. All rights reserved. Reprinted, with permission, from Wang et al. (2013)

Thus, the goal of the optimization is to obtain a contact configuration of the graphic tool that minimizes the deformation energy of the spring connecting it and the haptic tool.

Different parameter values in the matrix **G** will lead to different configurations for the graphic tool. This could be implemented intentionally to simulate different effects in non-realistic simulation scenarios such as computer games.

The flowchart of the configuration-based optimization approach to 6-DoF haptic rendering is shown in Fig. 2.3.

After computing the configuration of the graphic tool, the 6D feedback force and torque can be derived using a function that reflects the change of the relative configuration between the graphic tool and the haptic tool at current time step

$$\mathbf{F}^t = F\left(\mathbf{q}^t_g, \mathbf{q}^t_h\right) \tag{2.5}$$

where detailed functions of the above model depend on the physical properties of the tool and the objects.

In order for the configuration-based optimization approach to work effectively and efficiently, the following issues need to be resolved:

1. How to represent the tool and the object with complex shapes? The geometric representation should support diverse shapes, fast collision detection of possible contacts and penetrations, and modeling versatile contact scenarios, especially when the free moving space of the tool is a non-convex space. This issue will be elaborated in Sect. 2.2.
2. How to model, construct, and identify the non-penetration contact constraints between the graphic tool and the objects? Both single-region and multi-region contacts need to be modeled. The collision detection algorithm should be efficient. This issue will be elaborated in Sect. 2.2.
3. How to solve the optimization problem robustly and efficiently? Since contacts always occur at the surface of objects, while the constraints formula is expressed

in terms of configuration variables. Efficient transformation between the Cartesian space description of contact constraints and the configuration space of the graphic tool is needed to ensure fast collision response of 1 kHz. This issue will be elaborated in Sect. 2.3.
4. How to compute the feedback force and torque and ensure stability? This issue will be elaborated in Sect. 2.4.

2.2 Object and Contact Modeling Using Sphere Trees

Object and contact modeling is the fundamental step to realize the framework shown in Fig. 2.3. How to represent or model the geometry of objects in a virtual environment directly affects the computation performance for collision detection and contact constraint identification. An object is most commonly represented as a polygonal mesh, where the boundary of the object is approximated by connected triangles. A hierarchy of simpler bounding volumes, such as rectangles or spheres, is often used to speed up collision detection, with the polygonal mesh model at the bottom of the hierarchy (such as the leaf level of a tree structure). An object can also be represented as a sphere tree, where the root sphere encloses the entire object, and its children are smaller spheres enclosing parts of the object, and so on, so that at the leaf level, the spheres are the smallest in the greatest number to provide more accurate approximation of an object. Sphere trees provide convenience in representing complex contact constraints involving multiple contact regions that polygonal meshes lack. In the following, we first describe in detail the difficulty of formulating contact constraints between objects in polygonal mesh models and then present contact constraint formulation between objects in sphere-tree models. Finally, we describe how a sphere tree can be constructed.

2.2.1 Difficulties of Using Polygonal Meshes to Model Objects

Depending on if the space that the tool can freely move around, i.e., the free space, is convex or non-convex, contact constraints have to be formulated differently.

2.2.1.1 Convex Free Space

For the interaction between a convex-shaped tool and arbitrary-shaped objects, we need to identify different pairs of contact primitives between the graphic tool and the object. If mesh representations are used for both the tool and the object, the contact type in each contact region can be described by pairs of contacting geometric primitives, i.e., vertex, edge, and triangle face on each mesh.

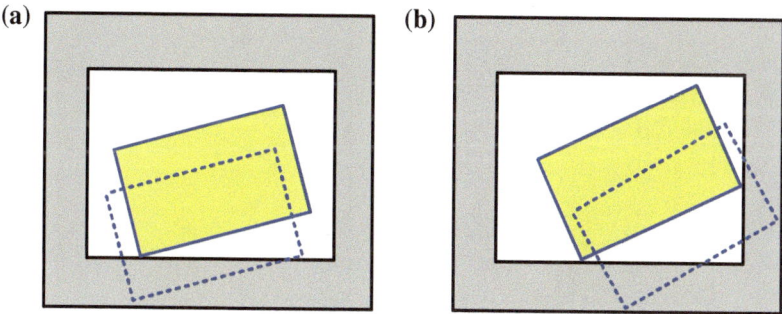

Fig. 2.4 Free space of the tool is a convex space. **a** Single-region contact. **b** Multi-region contacts

For the scenario in Fig. 2.4, a rectangular shaft in blue is moving within the interior volume of a rectangular hole, while the free space of the tool is a convex space. The dashed shaft represents the haptic tool, and the solid shaft represents the graphic tool. Denote a vertex of the tool as p_i, $i = 1, \ldots, 4$, and an unilateral surface constraint of the hole as F_i. A set of simultaneous point-face non-penetration constraints can be used for Eq. (2.1) in general, whether the contact is of a single region or multiple regions.

For the vertex-face contact type (V-F type), the constraint in Eq. (2.1) can be expressed as

$$g[p] \geq 0 \tag{2.6}$$

where $g[\cdot]$ represents a unilateral constraint formed by each triangle on the surface of the object. p refers to a boundary point or a vertex on the surface of the graphic tool. For each triangle on the surface of the object, we define a *constraint plane*

$$Ax + By + Cz + D = 0 \tag{2.7}$$

The normal of the plane is set as the normal of the triangle. Thus, we have

$$g[p] = \frac{(Ax + By + Cz + D)}{\sqrt{A^2 + B^2 + C^2}} \tag{2.8}$$

which is defined as the signed distance from the point $p(x, y, z)$ to the constraint plane.

We can express the constraint inequality for multiple point-plane constraints as follows:

$$g_i[p_j] \geq 0 \quad i, j = 1, \ldots, 4 \tag{2.9}$$

where g_i denotes unilateral constraints formed by the four planes (F_i) of the hole as shown in Fig. 2.5.

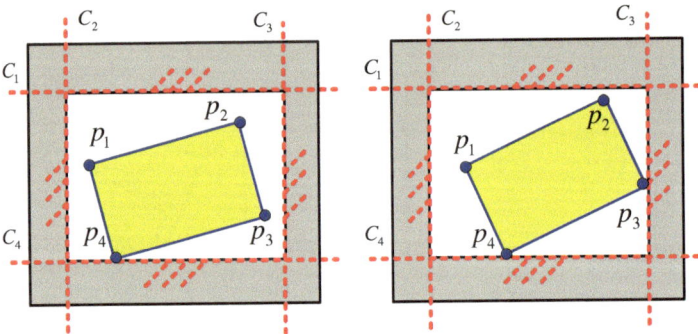

Fig. 2.5 Joint constraints functions (point-face constraints) for convex free space

These point-plane constraint inequalities consistently form an AND compound constraint, which defines the convex free space for the movement of the graphic tool.

2.2.1.2 Non-convex Free Space

For the scenarios shown in Fig. 2.6, the free space of the graphic tool is non-convex. In such a case, the related point-plane constraints do not consistently form an AND or an OR compound constraint. Figure 2.7 illustrates the constraints as forming an AND compound constraint and Fig. 2.8 illustrates the constraints as forming an OR compound constraint, but neither is correct for expressing the entire non-convex free space shown in Fig. 2.6. Thus, it is difficult to use a mesh model or a volume model to express the non-penetration constraints if the free space of the graphic tool is non-convex.

For a non-convex-shaped tool and an arbitrary object, such as shown in Fig. 2.9, the contact types are more diverse. It is even more difficult to use mesh-based object models to express the non-penetration constraints.

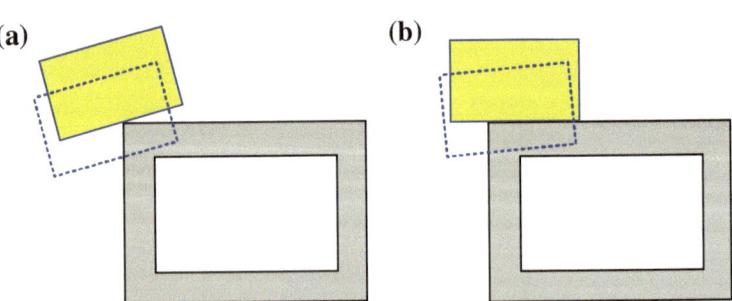

Fig. 2.6 Free space of the tool is a non-convex space. **a** Single-point contact. **b** Multi-point contacts

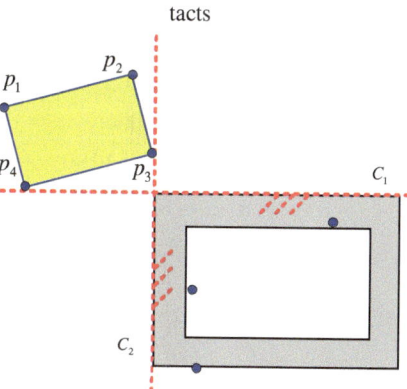

Fig. 2.7 Point-plane constraints form an AND compound constraint

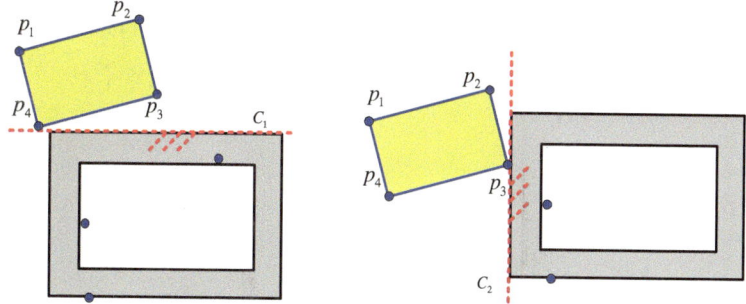

Fig. 2.8 Point-plane constraints form an OR compound constraint

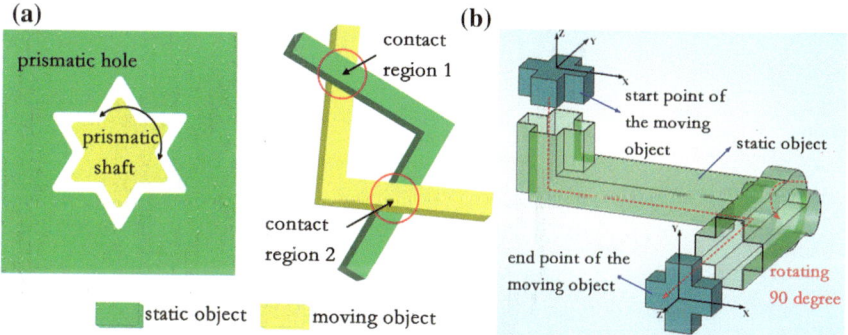

Fig. 2.9 Two interaction cases between two L-shaped objects: **a** edge–edge constraint; **b** combination constraints: edge–edge and point-face. *Note* the *red-shaded* region is the key challenge for constraint modeling

2.2.2 Contact Modeling Based on Sphere Trees

If a tool and an object are represented in sphere trees, no penetration between them means that there is no penetration between any sphere $s_T = (x_T, y_T, z_T, r_T)$ from the graphic tool's sphere tree and any sphere $s_O = (x_O, y_O, z_O, r_O)$ from the object's sphere tree, where (x, y, z, r) are the center and radius of a sphere in world coordinate system.

The non-penetration constraint can be written as follows:

$$(x_T - x_O)^2 + (y_T - y_O)^2 + (z_T - z_O)^2 \geq (r_T + r_O)^2 \tag{2.10}$$

Clearly, to model the non-penetration constraints between two objects, the sphere-tree representation provides a simple and uniform expression.

Since s_O is fixed in the virtual environment, and only s_T moves as the graphic tool moves, we can write (2.10) in a function form as follows:

$$C(x_T, y_T, z_T) \geq 0 \tag{2.11}$$

where x_T, y_T, and z_T are the functions of the graphic tool configuration $\mathbf{q}_g^t$, which are non-linear. We can establish their mapping using the following coordinate transformation.

First, we use L to represent the local coordinate system of the graphic tool, and use W to represent the world coordinate system. Then, we can construct the relationship between L and W as follows:

$$\mathbf{x}_W = \mathbf{T}_L^W \cdot \mathbf{x}_L, \tag{2.12}$$

where $\mathbf{x}_W$ and $\mathbf{x}_L$ represent the homogeneous coordinates of a point in W and L, and this point can be any sphere center of the graphic tool. And

$$\mathbf{T}_L^W = \begin{bmatrix} \mathbf{R}_L^W & \mathbf{p}_L^W \\ 0 & 1 \end{bmatrix}, \tag{2.13}$$

where

$$\mathbf{R}_L^W = \begin{bmatrix} c\alpha_g^t c\beta_g^t & c\alpha_g^t s\beta_g^t s\gamma_g^t - s\alpha_g^t c\gamma_g^t & c\alpha_g^t s\beta_g^t c\gamma_g^t + s\alpha_g^t s\gamma_g^t \\ s\alpha_g^t c\beta_g^t & ps\alpha_g^t s\beta s\gamma_g^t + c\alpha_g^t c\gamma_g^t & s\alpha_g^t s\beta_g^t c\gamma_g^t - c\alpha_g^t s\gamma_g^t \\ -s\beta_g^t & c\beta_g^t s\gamma_g^t & c\beta_g^t c\gamma_g^t \end{bmatrix} \tag{2.14}$$

and

$$\mathbf{p}_L^W = \begin{bmatrix} x_g^t & y_g^t & z_g^t \end{bmatrix}^\mathsf{T}, \tag{2.15}$$

where $\mathbf{q}_g^t = \left[x_g^t, y_g^t, z_g^t, \gamma_g^t, \beta_g^t, \alpha_g^t\right]^{\mathrm{T}}$, and the orientation angle α_g^t, β_g^t, and γ_g^t refers to the rotation angle around axis z, axis y, and axis x of the world coordinate system.

Take Eq. (2.12) into (2.11), we can construct non-linear constraint inequalities described in terms of the configuration variable of the graphic tool, i.e., the constraints in C-space can be expressed as follows:

$$C_i\left(x_T\left(q_g^t\right), y_T\left(q_g^t\right), z_T\left(q_g^t\right)\right) \geq 0 \quad i = 1, \ldots, N, \qquad (2.16)$$

where N refers to the number of contact sphere pairs.

To speed up the optimization, we linearize the above constraint using Taylor expansion. We will explain this linear model in Sect. 2.4.

2.2.3 Construction of Sphere Trees

Different sphere-tree models have been used in some existing penalty-based methods for efficient collision detection. Ruffaldi et al. proposed an implicit sphere tree to compute collision detection (Ruffaldi et al. 2006). They also introduced an adapted impulse-based method to solve the collision response. Weller et al. proposed a novel data structure, called inner sphere tree (IST), based on which they could easily handle distance-query and compute penetration volume (Weller and Zachmann 2009). Furthermore, they proposed a penalty-based collision response scheme which exploited penetration volumes to provide continuous force and torque.

Hubbard introduced a medial-axis-based sphere-tree model (Hubbard 1996), which is one of the best sphere-tree approaches to approximate an object with spheres. The key element of his algorithm is to find the skeleton of an object using a Voronoi diagram and create spheres from the skeleton to give a tight fit to objects. Bradshaw and O'Sullivan (Bradshaw and Sullivan 2004) have proposed several other algorithms (Merge, Burst, Expand, and Spawn) based on Hubbard's model and have extended all these algorithms in an adaptive manner. Compared to other sphere-tree models, the medial-axis sphere tree can not only approximate an object with a relatively small number of spheres, but can also achieve time-critical collision detection easily.

We use Bradshaw's sphere tree construction tool-kit (Bradshaw 2003), which contains a number of different algorithms to construct sphere trees of 3D objects. In particular, we use his combined algorithm to generate sphere trees with at most eight children per node. The combined algorithm allows the use of different sphere reduction algorithms in conjunction and chooses the one that result in the best fit. The sphere trees of a bunny and a dragon are shown in Fig. 2.10. The sphere trees of a dental probe and a tooth are shown in Fig. 2.11.

For each level in the sphere tree, we can compute the geometric error between the sphere models and the triangle-mesh representations. In Sect. 2.5.1, we use two

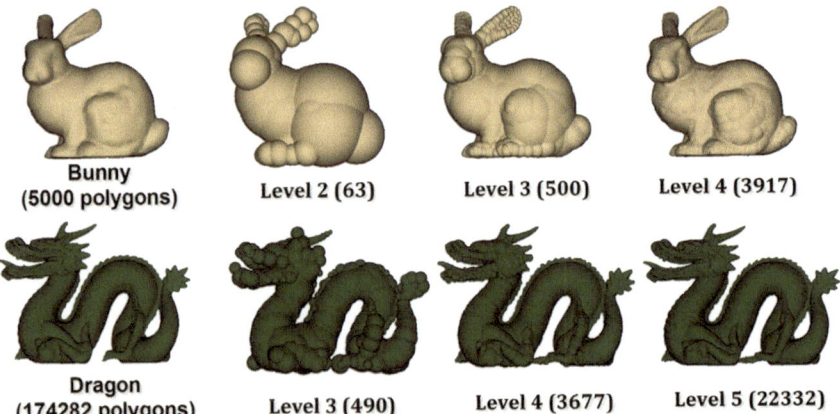

Fig. 2.10 Sphere trees generated for a bunny and a dragon. The *left* most figures are the polygonal models, and the rest are different levels of sphere trees. During implementation, level 4 (3917 spheres) is adopted to approximate a bunny and level 5 (22,332) is used for a more detailed dragon. Copyright © IEEE. All rights reserved. Reprinted, with permission, from Wang et al. (2013)

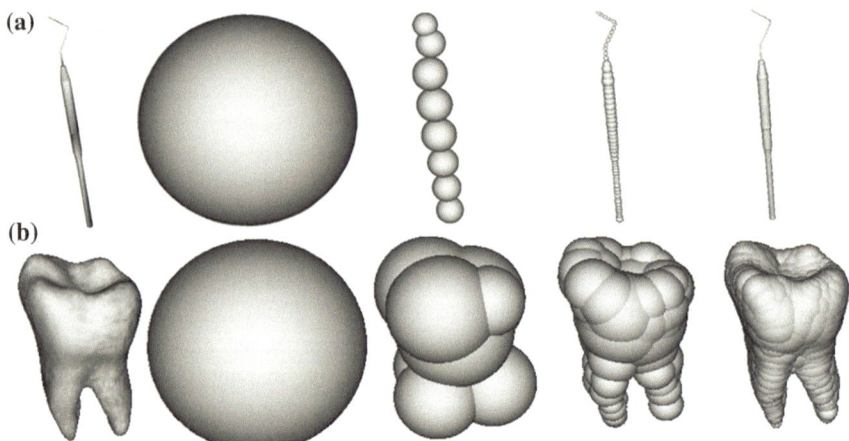

Fig. 2.11 Sphere trees generated for a dental probe and a tooth using the combined algorithm, **a** and **b** are models of the probe and tooth. The *left* most figures show the polygonal models, and the rest of the figures show the sphere trees varying from level 0 to level 3. Copyright © IEEE. All rights reserved. Reprinted, with permission, from Zhang et al. (2011)

benchmark objects to analyze the influence of the geometric error on the force-feedback fidelity.

Furthermore, physical parameters can be attached to each sphere, including stiffness, friction coefficient, and Young's Modulus.

2.3 Collision Response

We now describe how to compute the collision response when a tool interacts with objects in a virtual environment, using the configuration-based optimization approach and sphere-tree representation of the tool and objects. The computational pipeline is illustrated in following sub-sections.

2.3.1 Non-linear Transform from C-Space to Cartesian Space

In the optimization model shown in Eq. (2.1), the objective function is described in the configuration space (C-Space) of the tool, while the constraint functions are described in the Cartesian work space. There exists a non-linear mapping (in terms of trigonometric functions) between the two spaces because of the orientation components in the configuration of the graphic tool.

The coordinate transform between configuration variable $\mathbf{q}_g$ and Cartesian coordinate of the center of each sphere can be derived as Eq. (2.12).

2.3.2 Linearized Model of the Non-penetration Constraints

For fast computation, it is necessary to approximate the highly non-linear constraint for non-penetration with a linear model while keeping errors small for the solution variable (i.e., the configuration of the graphical tool). As changes of position and orientation of the haptic device are small between adjacent time steps, Taylor expansion is a suitable means for approximating the non-linear constraint.

We use a first-order Taylor expansion:

$$
\begin{aligned}
&C_i\left(x_T\left(\mathbf{q}_g^t\right), y_T\left(\mathbf{q}_g^t\right), z_T\left(\mathbf{q}_g^t\right)\right) \\
&\approx C_i\left(x_T\left(\mathbf{q}_g^{t-1}\right), y_T\left(\mathbf{q}_g^{t-1}\right), z_T\left(\mathbf{q}_g^{t-1}\right)\right) \\
&\quad + \sum_{j=1}^{6}\left[\frac{\partial C_i}{\partial x_T}\frac{\partial x_T}{\partial(\mathbf{q}_g^t)_j} + \frac{\partial C_i}{\partial y_T}\frac{\partial y_T}{\partial(\mathbf{q}_g^t)_j} + \frac{\partial C_i}{\partial z_T}\frac{\partial z_T}{\partial(\mathbf{q}_g^t)_j}\right] \cdot \left[(\mathbf{q}_g^t)_j - (\mathbf{q}_g^{t-1})_j\right]
\end{aligned}
\tag{2.17}
$$

The error caused by Taylor expansion is smaller than $O\left(\left(\Delta\mathbf{q}_g^t\right)^2\right)$. Because the change of configuration in one update cycle is relatively small, this error is imperceptible. Based on the preliminary test using a Phantom Premium device, the maximal velocity of the haptic tool in typical haptic operations is about 0.5 m/s. Because the update rate of the simulation is 1 kHz, the maximal displacement of the haptic tool is about 0.5 mm in each cycle. Therefore, the displacement error caused by Taylor expansion is about 0.25 mm, which is far smaller than the size of the simulated objects (e.g., bunny and dragon). Note that the error from Taylor expansion is not related to the size of a sphere in the sphere-tree model. Furthermore, the truncation error will not accumulate along a sliding motion of the tool, because the configuration of the haptic tool in each simulation cycle will automatically reset the possible error of the graphic tool in the cycle.

2.3.3 Active Set Method for Solving Constrained Optimization

As described earlier, the objective function of the optimization model is quadratic, and we linearize the contact constraints via Taylor expansion. Thus, this becomes a typical quadratic programming (QP) problem. Furthermore, in the optimization model shown in Eq. (2.1), all the constraint functions are inequalities instead of equations. We need to find an efficient way to solve this problem. This problem is solved using the active set method (Nocedal and Wright 2006), which is generally considered as one of the most effective methods for small- to medium-scale QP problems, performs well in the algorithm.

The method can always find the optimal solution in a finite number of iterations, but a good start can shorten this process greatly. The variation of the graphic tool's configuration from one rendering cycle to the next is small due to the high update rates of rendering. In other words, the two optimal points (in the C-space of the graphic tool) are near each other; therefore, the optimal point in the previous loop can be a good initial point for iteration in the current loop. We choose $\mathbf{q}_g^{t-1}$ as the initial solution point $\mathbf{q}_0$. Note that the initial solution/condition should be selected without causing penetration between the graphic tool and the object.

Using the active set method, the original optimization problem in Eqs. (2.1) and (2.3) can be transformed to the following form

$$\begin{cases} \min : f(\mathbf{q}) = \frac{1}{2}\mathbf{q}^T M \mathbf{q} \\ \text{s.t.} \quad J\mathbf{q} \geq -J\mathbf{q}_h^t \end{cases} \tag{2.18}$$

where J is formulated by Eq. (2.17), and

$$\mathbf{q} = \left(\mathbf{q}_g^t - \mathbf{q}_h^t\right) \tag{2.19}$$

Next, we use the active set method to find a step from one iteration to the next by solving a quadratic subproblem in which some of the constraints in (2.17) are imposed as equalities. Our algorithm continues to iterate in this manner until it reaches $\mathbf{q}_g^t$ that minimizes the quadratic objective function.

The main procedure of the active set method can be summarized as follows:

1. Find a feasible starting point;
2. Repeat until "optimal enough";

   (a) solve the equality problem defined by the active set (approximately);
   (b) compute the Lagrange multipliers of the active set;
   (c) remove a subset of the constraints with negative Lagrange multipliers;
   (d) search for infeasible constraints;
   (e) end repeat;

One limitation of the above active set solving method lies in the initial solution problem. Normally, we need to always maintain the initial solution in each simulation loop locating in the free space. As the initial solution is adopted as the solution of the graphic tool in the previous simulation cycle, the computation error by the truncation may produce small penetration. This issue may lead to the non-convergence of the iteration and needs to be further analyzed.

2.3.4 Contact Constraint-prediction Algorithm (CCP)

An important problem is how to find the contact constraints that the optimized configuration of the graphic tool has to satisfy, i.e., pairs of intersecting tool-object spheres. In previous work (Zhang et al. 2011), collision detection between the haptic tool and surrounding objects is carried out to obtain possible non-penetration and contact constraints. This conventional method leads to a large number of constraints when the shapes of the haptic tool and the object are complex. Furthermore, some constraints can be missed when the haptic tool is fully penetrated into objects. As shown in Fig. 2.12, if we use the pairs of intersecting spheres between the haptic tool and the object at time t to form the contact constraints for the configuration of the graphic tool at time t, the optimized (i.e., minimum energy) configuration of the graphic tool may not be penetration-free, that is, the contact constraints used in the optimization are not sufficient. To obtain sufficient number of constraints, the brute force method will be to form a non-penetration constraint from every pair of tool-object spheres, regardless if they are intersecting or not, which is inefficient.

Therefore, we introduce a *contact constraint-prediction* (CCP) method, taking advantage of the graphic tool's contact configuration in the previous time instant $t - 1$ and the motion coherency of the graphic tool. We use a proximity detection concept to enlarge the size of the graphic tool with a predefined threshold. The CCP method first decides the maximum distance δ that the tool can move in the interval between $t - 1$ and t based on the speed of the haptic tool. Next, each sphere of the graphic tool at $t - 1$ is enlarged by this threshold δ and the intersecting pairs of spheres between the enlarged graphic tool and the object are identified. These intersecting sphere pairs are used to form the contact constraints, as shown in Fig. 2.13.

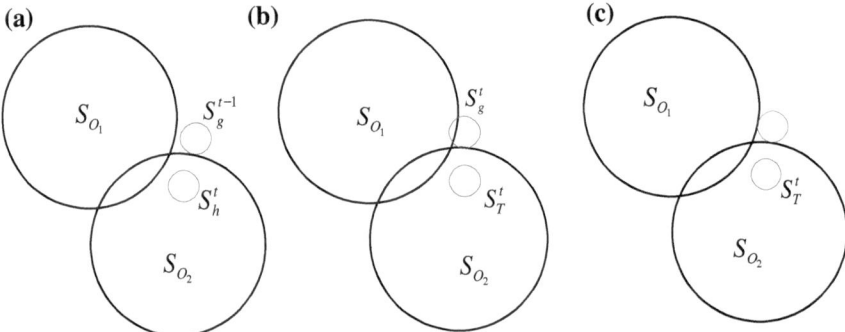

Fig. 2.12 Contact constraints from collision detection are not sufficient. **a** From collision detection, only the pair S_h^t and S_{O_2} is used to create a contact constraint. **b** S_g^t is the result of optimization with the single contact constraint from (**a**), which intersects O_1. **c** The desired result of optimization is shown as subject to both contact constraints from the two obstacle spheres. Copyright © IEEE. All rights reserved. Reprinted, with permission, from Wang et al. (2013)

It should be noted that a uniform δ is applied to all spheres of the graphic tool at time $t - 1$ based on the maximum distance the haptic tool can move in one haptic simulation cycle, which is corresponding to an estimated maximal velocity about 0.5 m/s.

Compared with the straightforward approach of detecting all pairs of intersecting spheres, the CCP method greatly reduces the number of constraints considered in the optimization without missing necessary constraints. We will demonstrate the benefit of the CCP method in the experiments in Sect. 2.6.

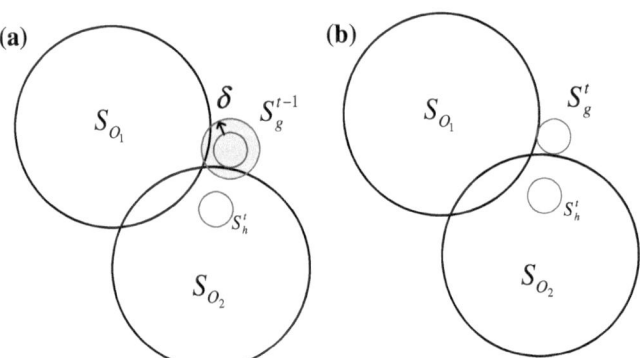

Fig. 2.13 CCP predicts all the possible contact constraints to ensure non-penetration of the graphic tool. **a** An enlarged sphere of the graphic tool at previous time step. **b** A desired position of the sphere of the graphic tool at current time step found using the CCP method. Copyright © IEEE. All rights reserved. Reprinted, with permission, from Wang et al. (2013)

2.4 Six-dimensional Force/Torque Simulation

After solving the configuration of the graphic tool, the next important step is to compute the contact force/torque exerted to the tool, and display them to the human user via the haptic device. Stability, fidelity, and fast response time are the main goals of force/torque computation and rendering models.

2.4.1 Existing Literature

Stability of haptic rendering is the most important criterion for designing a force/torque computation model. Early stability analysis in haptic rendering was focused on the problem of rendering stiff virtual walls. Several researchers reached the conclusion that high force update rates were necessary in order to achieve stable rendering (Colgate and Schenkel 1994; Brooks et al. 1990; Salcudean and Vlaar 1994). Intermediate representations (Adachi et al. 1995) were proposed to maximize the update rate of haptic rendering systems by performing a full update of the virtual environment at a low frequency (limited by computational resources and the complexity of the system) and using a simplified approximation for performing high-frequency updates of force feedback. Based on the z-width law of the haptic interface (Colgate and Brown 1994), to maintain the stability of a haptic simulation, the simulated stiffness of the virtual environment should be smaller than the maximum stiffness of the associated haptic device.

A number of techniques for 6-DoF haptic rendering follow the approach of direct rendering (Gregory et al. 2000; Kim et al. 2003; Johnson and Willemsen 2004). In direct rendering, the virtual tool follows rigidly the position of the haptic device, and contact forces are displayed directly. In this way, there is no need to simulate the dynamics of the virtual tool. However, guaranteeing stable and responsive display is a daunting task, as both the rotational impedance and the update rate of the haptic cycle may be highly varied. Small contact stiffness values result in large, visually perceptible interpenetrations, while large contact stiffness values may induce instabilities when the update rate of collision detection drops.

Colgate et al. (1995) proposed a multidimensional viscoelastic virtual coupling for stable interaction with non-linear virtual environments. If the implementation of the virtual environment is guaranteed to be discrete-time passive, the design of a stable display is reduced to appropriate tuning of the parameters of the coupling. Colgate and Schenkel (1994) pointed out that one way of obtaining a discrete-time passive virtual environment was to implicitly integrate a continuous-time passive system. Adams and Hannaford (1998) extended the concept of virtual coupling by providing a unifying framework for impedance and admittance displays.

Several techniques for 6-DoF haptic rendering combine virtual coupling with rigid body simulation of the virtual tool (McNeely et al. 1999; Chang and Colgate 1997; Berkelman 1999; Ruspini and Khatib 2000). Wan and McNeely (2003) instead employed a quasi-static simulation of the virtual tool. As mentioned before, the transparency of haptic display through virtual coupling may degrade with a slow simulation update rate or with a large virtual tool mass.

Researchers have also explored more flexible ways of ensuring system stability or passivity. Miller et al. (2000) extended Colgate's passivity analysis techniques by relaxing the requirement of passive virtual environments but enforcing cyclo-passivity of the complete system. Hannaford et al. (2002) investigated the use of passivity observers and passivity controllers. Mahvash and Hayward (2005) derived conditions for the passivity of a virtual environment where continuous-time passive local force models were activated sequentially.

2.4.2 Force/Torque Computation for Frictionless Contacts

Once the configuration of the graphic tool for time step t is obtained through configuration-based optimization, we use the following model to compute the contact force:

$$\mathbf{F}^t = [F_x, F_y, F_z]^T = G_t[x_g^t - x_h^t, y_g^t - y_h^t, z_g^t - z_h^t]^\mathrm{T} \qquad (2.20)$$

where G_t is a diagonal 3×3 stiffness matrix with k_t in the diagonal, where k_t denotes the translational spring stiffness between the graphic tool and the haptic tool.

Note that the virtual spring between the haptic tool and the graphic tool only takes effect when the haptic tool is partially or fully immersed into a virtual object, i.e., when it moves in the *constrained space*. The haptic tool and the graphic tool simply collocate if the haptic tool is in the *free space*. Thus, this method is different from the conventional "virtual coupling" method and avoids its damping effect.

The computed torque is with respect to a fixed world coordinate system W, and thus the torque computation procedure involves several coordinate transformations in a haptic simulation time step t, as described below:

1. Obtain the homogeneous transformation matrix of the haptic tool and its Euler angles with respect to the W.
2. Obtain the Euler angles of the graphic tool with respect to the W based on the optimized $\mathbf{q}_g^t$ of Eq. (2.18).
3. Compute the relative rotation matrix from the local coordinate frame of the graphic tool to the local coordinate frame of the haptic tool, using the following equation

$$_h^g\mathbf{R} = \mathbf{R}_L^{-1}(\mathbf{q}_h^t)\mathbf{R}_L(\mathbf{q}_g^t) \qquad (2.21)$$

where the rotation matrix $\mathbf{R}_L$ is defined as

$$\mathbf{R}_L(\mathbf{q}^t) = \begin{bmatrix} c\alpha^t c\beta^t & c\alpha^t s\beta^t s\gamma^t - s\alpha^t c\gamma^t & c\alpha^t s\beta^t c\gamma^t + s\alpha^t s\gamma^t \\ s\alpha^t c\beta^t & s\alpha^t s\beta^t s\gamma^t + c\alpha^t c\gamma^t & s\alpha^t s\beta^t c\gamma^t - c\alpha^t s\gamma^t \\ -s\beta^t & c\beta^t s\gamma^t & c\beta^t c\gamma^t \end{bmatrix} \tag{2.22}$$

4. Compute the torque with respect to the local coordinate frame of the haptic tool, by the following equation

$$\mathbf{T}_L^t = [T_z, T_y, T_x]_L^T = G_r \mathbf{\Psi}\left({}_h^g\mathbf{R}\right)_{3\times1} \tag{2.23}$$

where G_r is a diagonal 3×3 stiffness matrix with k_r in the diagonal, where k_r denotes the rotational spring stiffness between the graphic tool and the haptic tool. $\mathbf{\Psi}(\cdot)$ denotes a function that computes the three Euler angles $(\alpha^t, \beta^t, \gamma^t)^T$ based on a given rotation matrix as shown in Eq. (2.22). Detailed derivation of this function can be found in (Craig 1989).

5. Compute the torque with respect to W by the following equation

$$\mathbf{T}_W^t = [T_z, T_y, T_x]_W^T = G_r [\mathbf{R}_L(\mathbf{q}_h^t)]_{3\times3} \mathbf{\Psi}^t\left({}_h^g\mathbf{R}\right) \tag{2.24}$$

Note that in the above method, the continuity of the force/torque relies on the continuity of the graphic tool's configuration from one time step to the next. Given that the configuration of the haptic tool $\mathbf{q}_h^t$ continuously changes, and that the configuration of the graphic tool is computed as the local optimum solution that maintains a minimum Euler distance in six dimensions to the configuration of the haptic tool, the configuration of the graphic tool also changes continuously. Therefore, the computed force/torque changes continuously, and the stability of the haptic rendering can be maintained. The experiments introduced in the next section also empirically validate the stability of force/torque computation.

2.4.3 Force/Torque Computation for Frictional Contacts

In order to simulate subtle feeling of multi-region frictional contacts, we introduce the following extensions to the configuration-based method for friction simulation.

Given the known configuration of the graphic tool at the previous and current time steps $t - 1$ and t, and the known haptic configuration of the tool at t, we can compute the following distance vectors

$$\begin{cases} \delta\mathbf{p}_n^t = \mathbf{p}_{g\_eq}^t - \mathbf{p}_{h\_eq}^t \\ \delta\mathbf{p}_c^t = \mathbf{p}_{g\_eq}^{t-1} - \mathbf{p}_{g\_eq}^t \end{cases} \tag{2.25}$$

where $\delta\mathbf{p}_n^t$ and $\delta\mathbf{p}_c^t$ are two intermediate parameters denoting the average moving distance of several sampling points on the haptic tool, and

$$\mathbf{p}^t_{g\_eq} = \frac{1}{N} \sum_{k=1}^{N} \mathbf{p}^t_{g\_k} \qquad (2.26)$$

denotes the equivalent contact point of N contact points on the graphic tool at t. $\mathbf{p}^{t-1}_{g\_eq}$ denotes the corresponding point at time $t-1$. $\mathbf{p}^t_{h\_eq}$ denotes the corresponding point on the surface of the haptic tool at t.

In Eq. (2.25), we need to compute the corresponding point $\mathbf{p}^t_{h\_eq}$ of $\mathbf{p}^t_{g\_eq}$ by the following

$$\mathbf{p}^t_{h\_eq} = {}^W_h T \cdot \left[\left({}^W_g T \right)^{-1} \cdot \mathbf{p}^t_{g\_eq} \right] \qquad (2.27)$$

where ${}^W_h T$ and ${}^W_g T$ denote the homogeneous transformation matrix of the haptic tool and that of the graphic tool with respect to the world frame, respectively.

Since the tool and objects are modeled by sphere trees, the contact points in Eq. (2.26) can be found from intersecting spheres between the graphic tool and the objects after $\mathbf{q}^t_g$ is obtained from optimization.

We can compute the contact force signal (combining both normal and friction forces) as follows:

$$\mathbf{F}_{3*1} = k_e \left(\mathbf{p}^*_g - \mathbf{p}^t_h \right) \qquad (2.28)$$

where k_e denotes the maximum stiffness of the associated haptic device, $\mathbf{p}^t_h$ denotes the 3×1 position vector in $\mathbf{q}^t_h$ (i.e., the 6D configuration vector of the haptic tool), and

$$\mathbf{p}^*_g = \mathbf{p}^{t-1}_g + (1 - k_\alpha) \left(\mathbf{p}^t_g - \mathbf{p}^{t-1}_g \right) \qquad (2.29)$$

where $\mathbf{p}^t_g$ denotes the position vector of $\mathbf{q}^t_g$ (the 6D configuration vector of the graphic tool). Also

$$\begin{cases} k_\alpha = 1 & (\|\delta \mathbf{p}^t_c\|/\|\delta \mathbf{p}^t_n\|) \leq \mu_s \\ k_\alpha = \mu_d/\mu_s & (\|\delta \mathbf{p}^t_c\|/\|\delta \mathbf{p}^t_n\|) > \mu_s \end{cases} \qquad (2.30)$$

where μ_s denotes the static friction coefficient, which defines a coulomb friction cone (shown in Fig. 2.14) (Duriez et al. 2006). μ_d denotes the dynamic friction coefficient.

Similarly, we can derive the contact torque (combining torques from both normal and friction forces) as follows:

$$\boldsymbol{\tau}_{3*1} = k_\tau \left(\boldsymbol{\Phi}^*_g - \boldsymbol{\Phi}^t_h \right) \qquad (2.31)$$

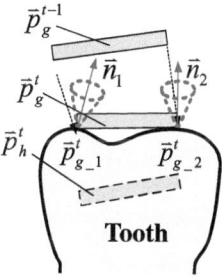

Fig. 2.14 Multi-point contacts, friction cones (*in blue*), and configurations involved in computing frictional contact force/torque. Copyright © IEEE. All rights reserved. Reprinted, with permission, from Wang et al. (2012a, b, c)

where k_τ denotes the maximum rotational stiffness of the associated haptic device. $\mathbf{\Phi}_h^t$ denotes the 3×1 vector of Euler angles in $\mathbf{q}_h^t$. $\mathbf{\Phi}_g^*$ denotes the vector of updated Euler angles after considering friction effect, which is computed by interpolating the general rotation angle as in the following steps:

1. The relative motion of the graphic tool from $t - 1$ to t can be derived and expressed by a homogeneous transformation matrix ${}_g^W T^t \cdot \left[\left({}_g^W T^{t-1} \right)^{-1} \right]$.

2. Using the general rotation matrix (Craig 1989), the equivalent general rotation axis $\bar{a}$ and the rotation angle θ_g can be obtained.

3. Let p be the contact point on the graphic tool that is closest to the rotation axis **a** at time t. Let the distance between p and **a** be r. Then define

$$\begin{cases} k_\theta = 1 & \left(r \cdot \theta_g / \| \delta \mathbf{p}_n^t \| \right) \leq \mu_s \\ k_\theta = \mu_d / \mu_s & \left(r \cdot \theta_g / \| \delta \mathbf{p}_n^t \| \right) > \mu_s \end{cases} \tag{2.32}$$

4. Next compute an interpolated rotation angle

$$\theta_g^* = (1 - k_\theta)\theta_g \tag{2.33}$$

5. Finally, compute the new Euler angles $\mathbf{\Phi}_g^*$ corresponding to the interpolated rotation angle.

   One important step to simulate static friction is to reset the configuration of the graphic tool to the configuration in the previous time step, i.e.,

$$\text{if } (k_\alpha = 1 \ \& \ k_\theta = 1), \text{ then } \quad \mathbf{q}_g^t = \mathbf{q}_g^{t-1} \tag{2.34}$$

This means that the graphic tool is stuck on the object's surface (i.e., neither translation nor rotation can occur). It should be noted that the above reset operation may occur for several consecutive steps. The number of the consecutive reset steps

N_{reset} is a function of the time that the human operator pauses the movement of the haptic tool and the speed of the operator's movement of the haptic tool.

Compared to the iterative solutions using Gauss–Seidel algorithm (Duriez et al. 2006), the mathematical model of the method is easier to be implemented by engineering developers.

2.5 Experimental Validation

In this section, we introduce a number of experiments using benchmark virtual environments to validate the configuration-based optimization approach to 6-DoF haptic simulation.

A Phantom Premium 3.0 6-DoF is used as the haptic device to display six-dimensional forces and torques (shown in Fig. 2.15). The specifications of the computer are as follows: Intel (R) Core (TM) 2 2.20 GHz, 2 GB memory, and X1550 Series Radeon graphical card. Note that no GPU is used in the implementation of the haptic simulation method.

Three metrics are considered in performance evaluation:

- Accuracy: When a contact occurs, there is no visual artifact (i.e., no interpenetration between the tool and the object), and no haptic artifact (i.e., no force felt when there is still a distance between the tool and the object, or no artificial friction and sticking when the tool moves along the object's surface).
- Stability: There is no big jump or vibration of the graphic tool between two adjacent simulation time steps, and the feedback force/torque is smooth and stable.

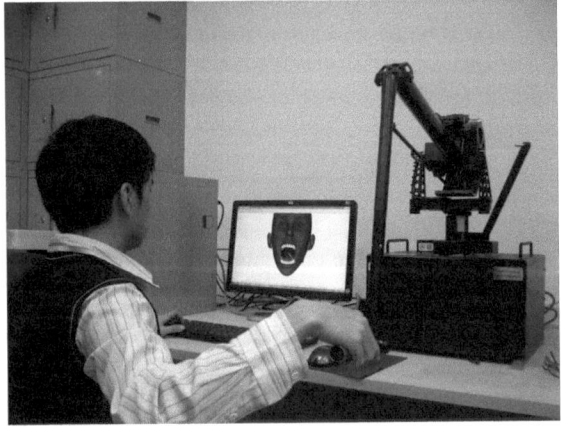

Fig. 2.15 Interaction with a Phantom Premium 3.0/6-DoF device. Copyright © IEEE. All rights reserved. Reprinted, with permission, from Wang et al. (2013)

- Computation update rate: the update rate is higher than 1 kHz under all contact scenarios (i.e., either a single contact or a multi-region contact case).

2.5.1 Results of Accuracy Analysis

We evaluate accuracy in terms of accuracy of the computed force/torque signal and accuracy of the configuration of the graphic tool.

Accuracy of the computed feedback force/torque signal is defined as the following error at any time step to measure the haptic realism of the target haptic rendering method:

$$\begin{cases} \delta_F = \left(\mathbf{F}_{gm} - \mathbf{F}_{tm}\right)/\left\|\mathbf{F}_{gm}\right\| \\ \delta_\tau = \left(\tau_{gm} - \tau_{tm}\right)/\left\|\tau_{gm}\right\| \end{cases} \tag{2.36}$$

where $\mathbf{F}_{gm}$ and $\mathbf{F}_{tm}$ indicates the three-dimensional force signal from a ground-truth model and a target model, respectively. τ_{gm} and τ_{tm} indicates the three-dimensional torque signal from the ground-truth model and the target model, respectively.

Similarly, the error of position and orientation signal of the graphic tool is defined to evaluate the visual realism of the target haptic rendering method:

$$\begin{cases} \delta_X = \left(\mathbf{x}_{gm} - \mathbf{x}_{tm}\right)/\left\|\mathbf{x}_{gm}\right\| \\ \delta_\theta = \left(\boldsymbol{\theta}_{gm} - \boldsymbol{\theta}_{tm}\right)/\left\|\boldsymbol{\theta}_{gm}\right\| \end{cases} \tag{2.37}$$

where $\mathbf{x}_{gm}$ and $\mathbf{x}_{tm}$ indicate the three-dimensional position signal of the graphic tool from the ground-truth model and the target model, respectively. $\boldsymbol{\theta}_{gm}$ and $\boldsymbol{\theta}_{tm}$ indicate the three-dimensional orientation angle of the graphic tool from the ground-truth model and the target model, respectively.

Based on those error definitions, the smaller of the error, the more accurate is the target haptic rendering algorithm.

The accuracy of simulation comes from the quadratic programming (QP) formulation. In contrast to previous methods, exact non-penetration between the graphic tool and the object can be maintained, without introducing numeric error caused by integration computations. Possible factors that can affect accuracy are the geometric error of using spheres to approximate an object's shape and the error incurred in using Taylor expansion. Our experiments quantify that those two factors have negligible effects so that the method can achieve high accuracy.

In each experiment, we use analytic functions to describe a ground-truth model, i.e., an object with analytic expression of its geometry. Specifically, an ellipsoid or a cube is used. A sphere with 5 mm radius is used as the tool. Different levels of sphere trees are utilized for the cube and the ellipsoid. The experiments are expected to show that the geometric error between a high-resolution sphere-tree model and the analytic model of the object is sufficiently small, so that the force

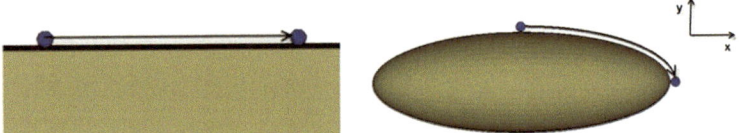

Fig. 2.16 Sliding trajectory of tool on the target object's surface is confined to the *x-y* plane. Copyright © IEEE. All rights reserved. Reprinted, with permission, from Wang et al. (2013)

error (computed by Eq. 2.36) between the sphere-tree model and the analytic model is also small enough, i.e., smaller than human's discriminating threshold.

Here, a user moves the haptic tool along a fixed path on the planar surface of the cube or the curved surface of the ellipsoid represented by a sphere tree (as shown in Fig. 2.16). The ideal force magnitude should be 1 N, and the ideal force direction should be along the surface normal.

In each case, we record samples of the contact force. The errors of contact force components between the ideal plane and the sphere-tree model are shown in Fig. 2.17. The horizontal axis denotes the simulation time step (in millisecond), and vertical axis denotes the error of force components computed by Eq. (2.36). Similarly, the errors of contact force components between the ideal ellipsoid and the sphere-tree model are shown in Fig. 2.18. It should be noted that the errors shown are only for force (not torque) along the x- and y-axes, because the predefined sliding trajectory is confined to the *x-y* plane, and there is no rotation.

From the figures, we can see clearly that the geometric error decreases as deeper sphere trees with more spheres are utilized. But the trend is not necessarily linear. Rigorous experiments or theoretical analysis is needed to determine the exact relationship between the error and the resolution of the sphere tree. For the cube model, the average force error under 4,096 spheres is at about 0.05. For the

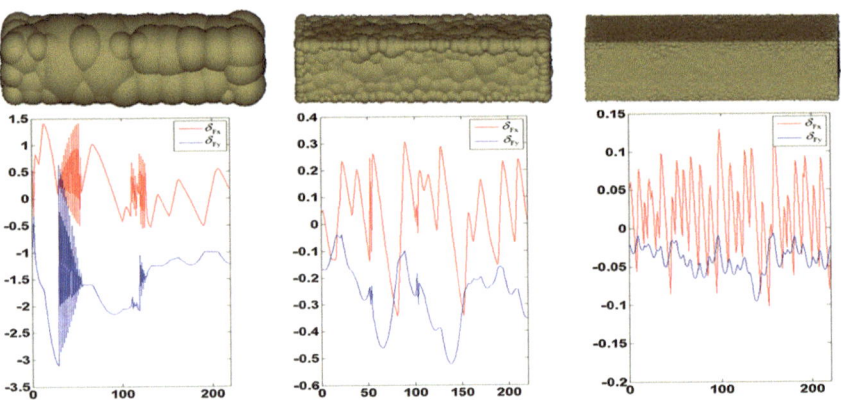

Fig. 2.17 Results of sliding along the cube (the number of spheres are 64, 512, and 4,096, respectively). Copyright © IEEE. All rights reserved. Reprinted, with permission, from Wang et al. (2013)

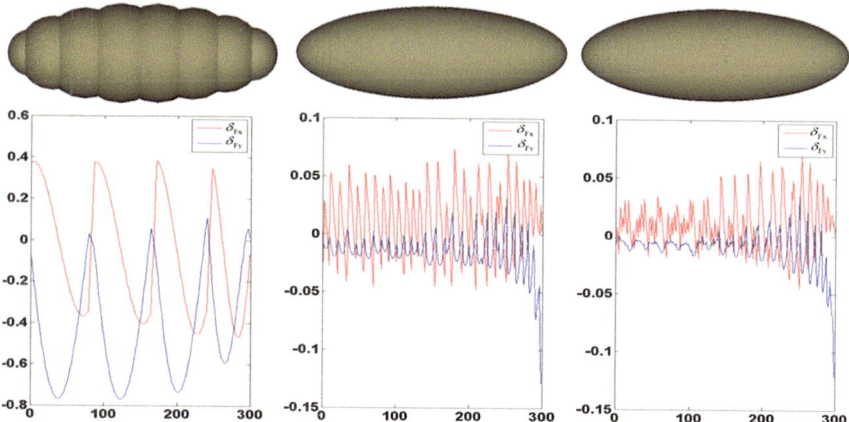

Fig. 2.18 Results of sliding along the ellipsoid (the number of spheres is 8, 64, and 512, respectively). Copyright © IEEE. All rights reserved. Reprinted, with permission, from Wang et al. (2013)

ellipsoid model, the maximum force error under 512 spheres is at about 0.05. Since human's Just Noticeable Difference (JND) for force magnitude is about 6–8 % (Jones et al. 1992; Pang et al. 1991; Tan et al. 1995), the average 5–10 % force errors in Figs. 2.20 and 2.21 suggest that the computed force/torque based on the sphere-tree models is accurate enough comparing to the computed force/torque based on the precise, analytic model of the tool, and object.

Using the sphere-tree model is more suitable for approximating curved objects than using polygonal mesh models. For a ground-truth model such as the ellipsoid, a sphere-tree model with a total of 64 spheres can achieve lower force error than a polygonal mesh with over 4,000 triangles. On the other hand, to simulate straight edges or flat surfaces, a finer sphere-tree model is required than simulating a curved object. As shown in Figs. 2.20 and 2.21, the sphere tree with a total of 4,096 spheres to simulate the cube yields a greater force error than the sphere tree with a total of 512 spheres to simulate the ellipsoid. The perceptual artifacts depend on the shape of the explored object.

As for the truncation error caused by the Taylor expansion, it is recorded in all the experiments, when the tool slides along the object's surface. The error values are always less than one micrometer. Hence, it is sufficiently small so that the corresponding visual artifact is negligible.

2.5.2 Stability and Update Rate—Exp 1: Dental Operations

In a dental operation, multi-region contacts often occur between the dental tool and oral tissues. Frequent contact switches occur when the tool moves or rotates within the small oral cavity or even in the narrow cavity of periodontal pocket. Two kinds of dental operations are simulated to validate the stability of the haptic rendering method.

2.5.2.1 Periodontal Operation

One important periodontal operation is a periodontal pocket examination through probing. Periodontal tissue consists of gingiva, periodontal ligament, alveolar bone, and root cementum. Figure 2.19 shows the comparison between healthy periodontium and inflamed periodontium. As shown in Fig. 2.19c, periodontal pocket is pathologically deepened gingival sulcus (the shallow crevice between free gingiva and tooth surface). Therefore, periodontal pocket probing is an important operation for periodontal diagnosis. During the examination, a dentist uses a periodontal probe to check the depth value of the pocket, and thus to determine the degree of the periodontal inflammation.

This operation involves frequent changes of multiple contacts and thus is ideal for checking the stability of the haptic simulation method. As shown in Fig. 2.20, first, the operator slides the probe on the tooth and gingiva until finding the location of inflammation. Then, the tool is manipulated into the sulcus between the tooth and gingiva for inspection. The quite even curves of the force and torque demonstrate the stability of the algorithm. Also, the operator cannot feel any vibration of the haptic device. After the twentieth second, we can see clear increase of the torque

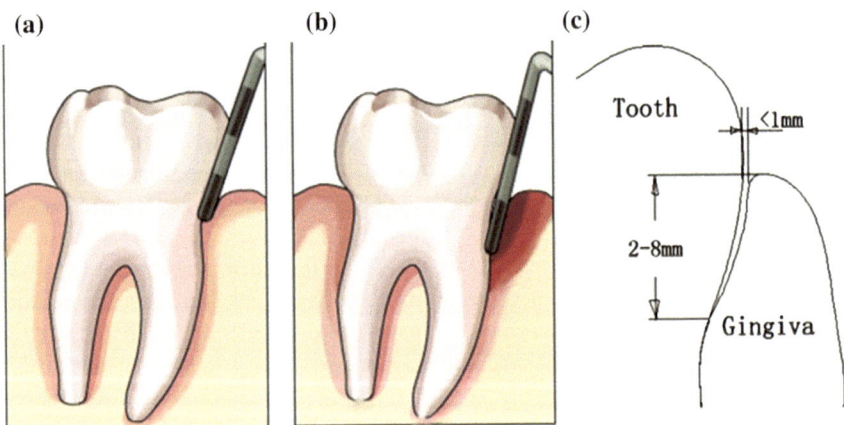

Fig. 2.19 A periodontal pocket. Copyright © IEEE. All rights reserved. **a** Healthy periodontium, **b** periodontium with inflammation, **c** illlustration of periodontal pocket. Reprinted, with permission, from Wang et al. (2013)

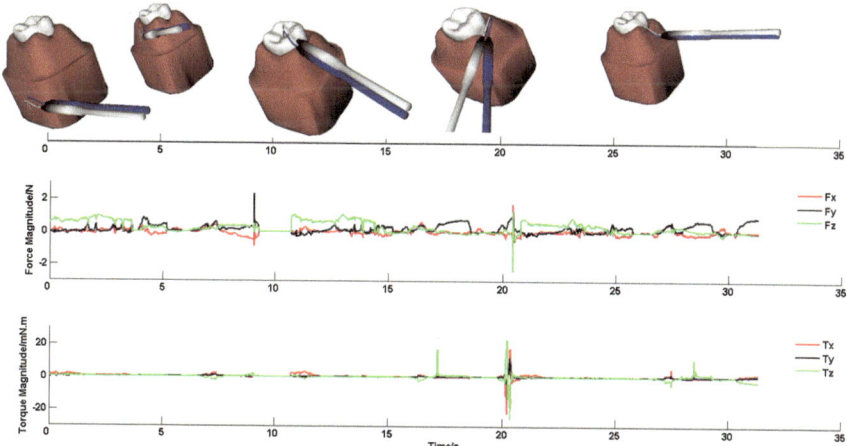

Fig. 2.20 Although the haptic tool (the *blue one*) has penetrated into the tooth/gingival, the graphic tool (the *white one*) remains just in contact. Copyright © IEEE. All rights reserved. Reprinted, with permission, from Wang et al. (2013)

signal when the operator voluntarily rotates the tool in the periodontal pocket. Non-penetration is maintained between the graphic tool (in white color) and the tissues (both tooth and gingiva). A video showing the process is available online from the book Website http://extras.springer.com/.

2.5.2.2   Oral Cavity Inspection

In oral cavity inspection, a dentist uses a dental probe to examine possible diseases such as angular stomatitis, glossitis, swollen or bleeding gums, and decayed teeth. For this task, a complete oral virtual environment is necessary to simulate the movement of the probe within the oral cavity, which includes all teeth, lower jaw, upper jaw, tongue, and cheek. Figure 2.21a shows the dental surgery simulator. From Table 2.1, we can see the total number of triangles and leaf-node spheres is rather large in this experiment.

This is a challenging task to simulate at the desired haptic rate (i.e., 1 kHz). We test the simulation in the following steps: (1) Examine every single tooth to see if any decay exists. (2) Examine the gingiva to see if it is inflamed (this manipulation will make the tool contact several parts in a mouth, and this is why the curves shown in Fig. 2.21 rise up after the twentieth second). The results shown in Fig. 2.21b demonstrate that the haptic rendering method is efficient and provides stable interaction. A video showing the process is available online from the book Website http://extras.springer.com/.

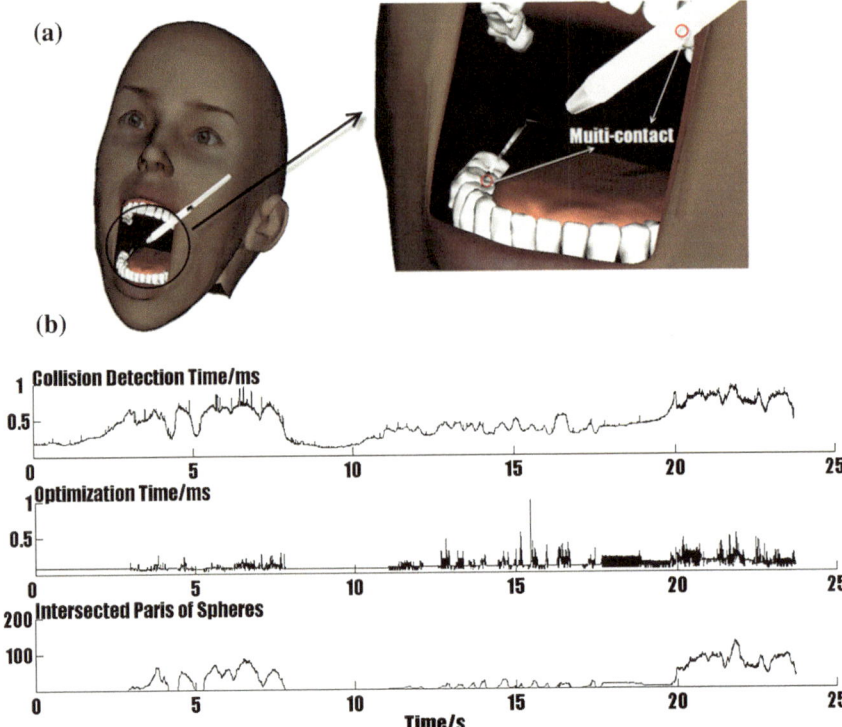

Fig. 2.21 Oral cavity inspection. Collision detection time and optimization time are related to the number of intersected spheres. The algorithm can still maintain haptic rates even though hundreds of spheres intersect. Copyright © IEEE. All rights reserved. Reprinted, with permission, from Wang et al. (2013). **a** Interaction scenario. **b** Time cost for collision detection and optimization

Table 2.1 Number of triangles and leaf-node spheres in the oral cavity inspection experiment

	Polygons	Leaf-node spheres
Probe	1,088	512
Face	3,138	512
Upper jaw	2,280	512
Lower jaw	2,280	512
Tongue	1,045	512
Teeth	28 × 4,000	28 × 512

2.5.3 Stability and Update Rate—Exp 2: Bunny Versus Bunny

In Fig. 2.22, the brown bunny represents the graphic tool, and the white bunny represents the static object. Both bunnies are modeled using level-4 sphere trees (with 3,076 spheres in each bunny).

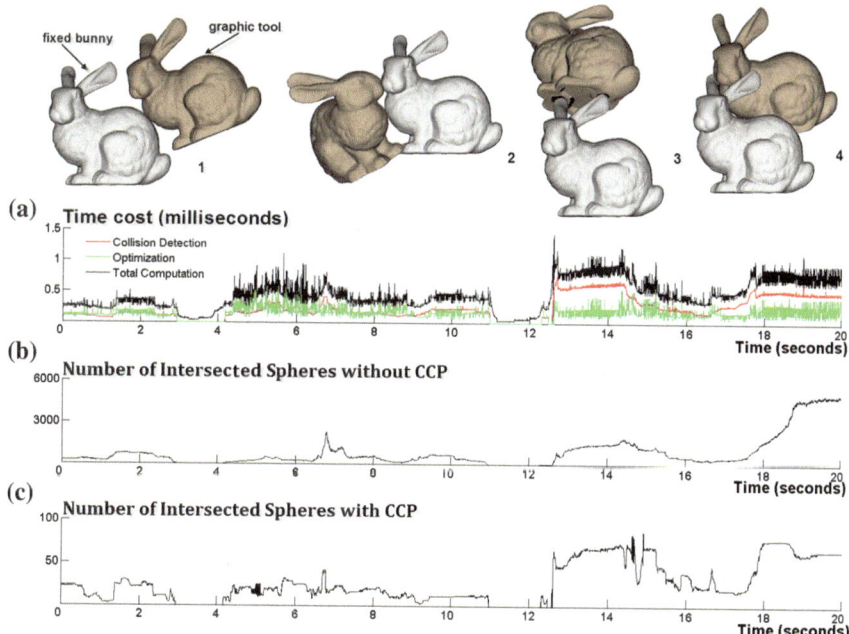

Fig. 2.22 Results of haptic interaction with the Stanford bunnies. *Top* four key states of the interaction (the *white bunny* is static). Plot **a** time cost of the main algorithms in the approach. Plot **b** the number of intersected spheres with conventional collision detection. Plot **c** the number of intersected spheres with CCP method. Copyright © IEEE. All rights reserved. Reprinted, with permission, from Wang et al. (2013)

During the simulation, several contact states are examined. In state 1, the brown bunny slides along the back of the white bunny to validate whether the method can simulate continuous contact between the tool and the object. In state 2, the brown bunny makes a multi-region contact (with three simultaneous contact regions). In state 3, the white bunny's two ears get into the holes at the bottom of the brown bunny for fine interaction. In state 4, the brown bunny is squeezed between the two ears of the white bunny. When the operator rotates the brown bunny and maintain the contacts with the two ears, frequent contact switches occur. The interaction is very stable.

The time efficiency is one of the most significant advantages of the method compared to existing constraint-based haptic rendering methods. Figure 2.22a shows the efficiency of the method. The time cost of the collision detection fluctuates when the contact state between the tool and the object changes. The maximum time cost of the collision detection is less than 1 ms. The time cost of the optimization is always less than 0.5 ms, benefiting from the selection of the initial iteration configuration. Therefore, the overall time cost in each simulation cycle is less than 1.5 ms, which meets the strict update rate for haptic rendering. Whereas for the same example of bunny-bunny interaction in (Ortega et al. 2007), the simulation time for the god-object is in the range of 1–15 ms. The method is more efficient because in (Ortega et al. 2007), the continuous collision detection (CCD) method relies on mesh–mesh collision check, which is more time intensive, and it is embedded in the main simulation loop.

Figure 2.22b indicates the number of intersected spheres using conventional discrete collision detection (i.e., checking intersection between the haptic tool and the virtual environment), and Fig. 2.22c indicates the number of intersected spheres using the CCP method. Comparing Fig. 2.22b with c, we can see that, CCP not only provides sufficient constraints, but also greatly reduces redundant constraints, which makes the algorithm more efficient.

2.5.4 Stability and Update Rate—Exp 3: Buddha Versus Dragon

Figure 2.23a shows a moving Buddha statue interacting with a static dragon statue. Both objects are modeled by a level-5 sphere tree consisting of more than 20,000 spheres. We compute a non-penetration configuration of the graphic tool (as shown in Fig. 2.23a) using different levels of the sphere tree (note that the Buddha and the dragon have the same level). Figure 2.23b shows a state where the Buddha and the

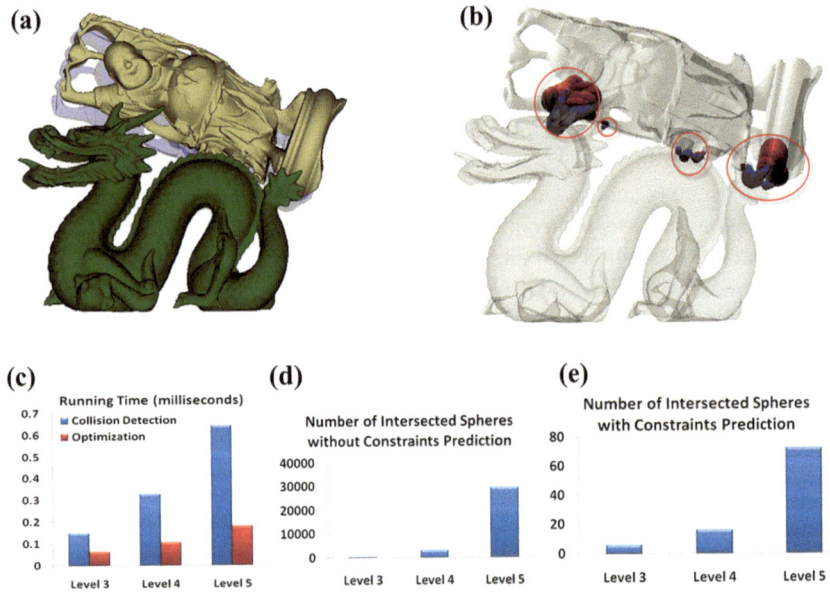

Fig. 2.23 Haptic simulation of multi-region contacts involving statues of a moving Buddha and a static dragon. Copyright © IEEE. All rights reserved. Reprinted, with permission, from Wang et al. (2013)

dragon have four simultaneous contact regions. From Fig. 2.23c, one can tell the time costs of the collision detection and optimization increase along with the increased levels. However, they are below 1 ms even for level 5 (with more than 20,000 spheres). Considering the contact distribution, the number of simultaneous contact pairs of spheres cannot be too large; otherwise the time cost for optimization will exceed 1 ms. From the experimental results, the maximum number of simultaneous pairs of intersecting spheres can be found to be about 500. As shown in Fig. 2.23d and e, the number of intersecting spheres after using the CCP algorithm is kept to a relatively low level.

The time costs for collision detection and optimization in one haptic simulation cycle are provided in Fig. 2.23. From this figure, we can see the time cost for collision detection is maintained within 1 ms even for highly detailed sphere trees (e.g., level-5 sphere trees for the dragon and Buddha). Using the CCP method, the number of intersecting sphere pairs is greatly reduced, and thus the time cost for optimization is greatly reduced.

2.5.5 Stability and Update Rate—Exp 4: Bunny Navigating

This experiment is carried out to further illustrate the performance of the method in a complex scenario with frequent contact switches.

A pipe system consists of 17 separate pipes, and a bunny is manipulated to move through it. We show four key states of the motion in Fig. 2.24, and the black curve indicates the path of the bunny's centroid. Because it is a complex, irregular virtual environment, a human operator is required to find a way out for the bunny relying on the force feedback. In this experiment, frequent contact switches occur, i.e., the number of the contact regions and the pairs of intersecting spheres in each region will change whenever small translation/rotation of the bunny occurs. The approach performs well for such a challenging task with frequent contact switches.

As shown in Fig. 2.24a, b, the force/torque magnitudes change frequently in all three directions. However, as shown in Fig. 2.24c, d, the difference between adjacent force/torque values is relatively small, i.e., the stability of the haptic manipulation is maintained. The update rate of the haptic simulation cycle is always over 1 kHz.

2.5.6 Validation of Friction Simulation

As shown in Fig. 2.25, the force signals (with or without friction effect) during two continuous movement segments are recorded. In the two segments, the operator moves the tool approximately along z-axis and x-axis, respectively. During each movement segment, multi-region contacts occur between the graphic tool and the tooth.

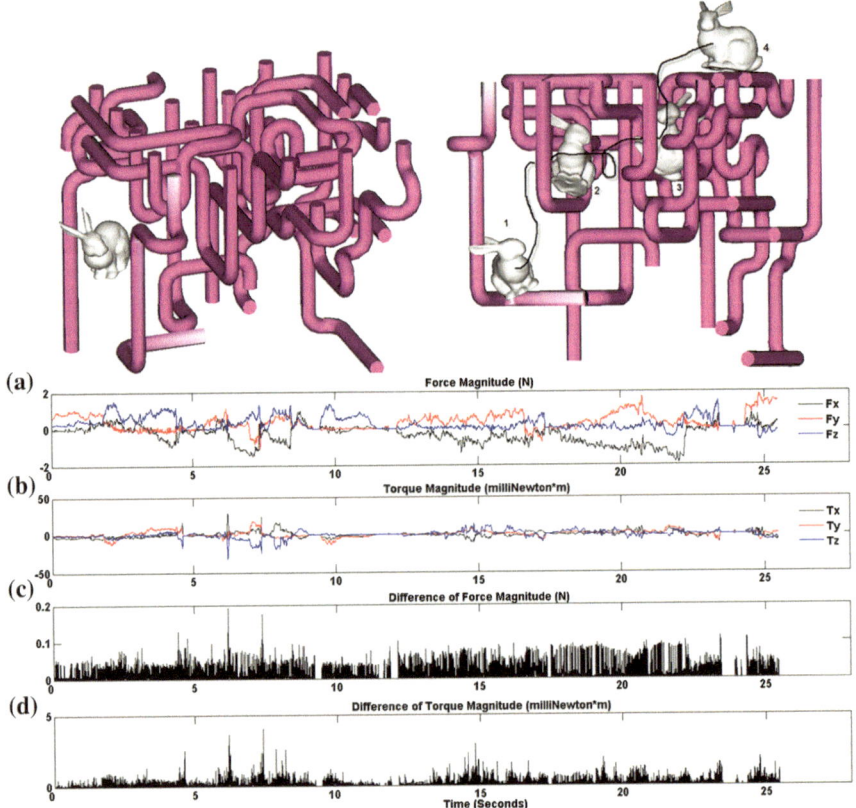

Fig. 2.24 A bunny through a pipe system to simulate frequent contact switches. State 2 indicates manipulating the bunny through U-shape pipes. From state 2 to state 3, the bunny is turned around to make its body get through first so that the bunny's ears can get through. Copyright © IEEE. All rights reserved. Reprinted, with permission, from Wang et al. (2013)

We can see that the component of the friction force is consistent with the movement direction of the haptic tool, i.e., the friction component along z-axis is significant in the first segment, and the friction component along x-axis is significant in the second segment. The static and dynamic friction coefficients were set as 0.8 and 0.3, respectively, in the force/torque curves.

Because the movements of the tool might not be strictly aligned with the z direction in the first segment, there is also an increase in the force along x-axis. During the second segment, the movement is sliding motion on top of the tooth.

Next, progressive change of friction force is validated. A circular area is set as a calculus. Because the scaler and the calculus are both modeled by sphere trees, multi-point contacts occur during the scaling process. As shown in Fig. 2.26, we can observe the decrease of friction force along y-axis during the removal process (as shown in video, the scratching movement of the scaler is mainly along the y-axis). As calculus removal proceeds, the number of reset steps becomes smaller,

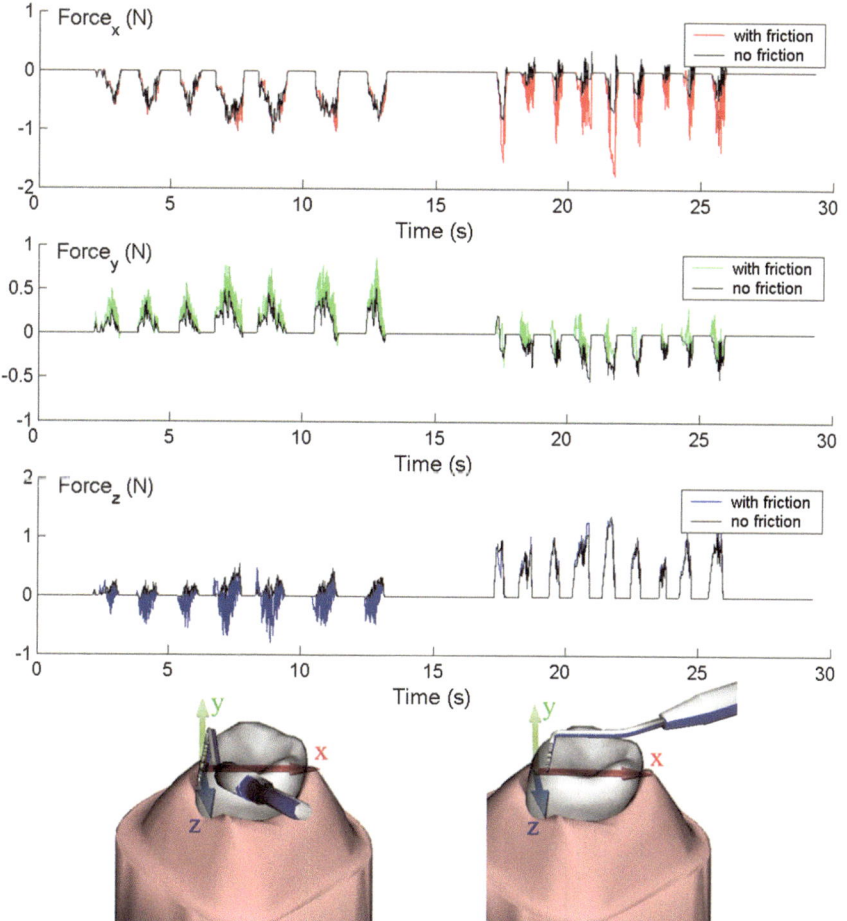

Fig. 2.25 Multi-contact force signals including both normal and friction effect during two movement segments. Copyright © IEEE. All rights reserved. Reprinted, with permission, from Wang et al. (2012a, b, c)

and the magnitude of friction force also becomes smaller. After the calculus was completely removed, no friction could be felt. A video showing the process is available online from the book Website http://extras.springer.com/.

The configuration-based simulation method for multi-contact friction is very efficient and can maintain 1 kHz update rate of the haptic loop without the use of special computation hardware such as GPU or multi-core. It can simulate progressive or changing friction effect conveniently and stably.

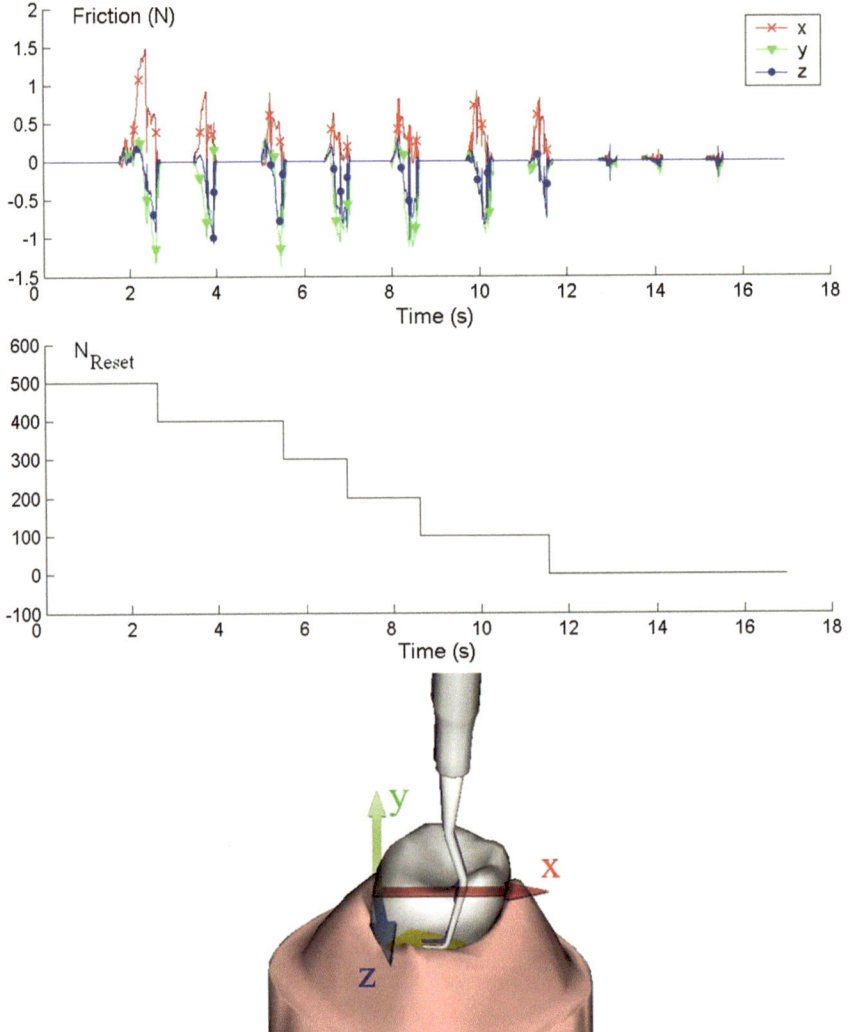

Fig. 2.26 Changing friction effects during calculus removal process by gradually change the value of the number of reset steps. Copyright © IEEE. All rights reserved. Reprinted, with permission, from Wang et al. (2012a, b, c)

2.6 Summary

In this Chapter, we have introduced an efficient constraint-based 6-DoF haptic rendering approach, i.e., a configuration-based optimization method based on sphere-tree representations of objects to compute the collision response for 6-DoF haptic rendering for fine manipulation.

This approach is a significant extension of the god-object method in three-dimensional space (Zilles and Salisbury 1995). With this approach, visual artifacts are avoided, i.e., no perceptible penetration or separation between the graphic tool and an object. The graphic tool is stopped at the surface of the object. Different from existing constraint-based haptic rendering methods, the method is efficient such that it computes collision response at the 1 kHz haptic rate. Hence, it can capture the subtle change of contact constraints in adjacent time steps and can simulate multi-region contacts stably when the tool moves within a narrow space.

The efficiency of the method is ensured by a number of techniques. Sphere trees are used to represent both the tool and objects and model non-penetration constraints between the graphic tool and arbitrary-shaped objects in a simple and uniform expression. In addition to taking advantage of quick intersection check between spheres, a contact constraint-prediction (CCP) method further ensures the correctness and computation efficiency of finding contact constraints. Taylor expansion is used to linearize the non-linear constraint inequalities to achieve and maintain high computation rate for the optimization while maintaining small truncation errors. Finally, feedback force/torque is computed without using virtual coupling.

Extensive experimental testing demonstrates that with this method, high accuracy, stability, and high update rates are achieved in some complex task environments with multi-region contacts and frequent contact switches.

For multi-region contacts involving large contact regions, the time cost of the optimization method will become a bottleneck because the number of the contact pairs can be very large. Even using CCP, the time cost will exceed 1 ms. This problem can be solved by eliminating some redundant constraints before doing optimization.

In the following chapters, this approach will be extended to simulating fine geometric features such as sharp edges, and to virtual environments consisting of both rigid and deformable objects.

References

Adachi Y, Kumano T, Ogino K (1995) Intermediate representation for stiff virtual objects. In: Proceedings of virtual reality annual international symposium, pp 203–210

Adams RJ, Hannaford B (1998) A two-port framework for the design of unconditionally stable haptic interfaces. In: Proceedings of IEEE/RSJ international conference on intelligent robots systems, pp 1254–1259

Berkelman PJ (1999) Tool-based haptic interaction with dynamic physical simulations using Lorentz magnetic levitation. PhD dissertation, Robotics Institution, Carnegie Mellon University, Pittsburgh, PA

Bradshaw G, Sullivan CO (2004) Adaptive medial-axis approximation for sphere-tree construction. ACM Trans Graph 23(1):1–26

Bradshaw G (2003) Sphere-tree construction toolkit. http://isg.cs.tcd.ie/spheretree/, Feb 2003

Brooks Jr FP, Ouh-Young M, Batter JJ, Kilpatrick PJ, Baskett F (eds) (1990) Project GROPE—haptic displays for scientific visualization. In: Proceedings of computer graphics (SIGGRAPH), Aug 1990, vol 24, pp 177–185

Chang B, Colgate JE (1997) Real-time impulse-based simulation of rigid body systems for haptic display. In: Proceedings of ASME Dynamic Systems and Control Division, pp 145–152

Colgate JE, Brown JM (1994) Factors affecting the Z-width of a haptic display. In: Proceedings of IEEE international conference on robotics and automation, Los Alamitos, CA, pp 3205–3210

Colgate JE, Schenkel GG (1994) Passivity of a class of sampled-data systems: application to haptic interfaces. In: Proceedings of American control conference, pp 3236–3240

Colgate JE, Stanley MC, Brown JM (1995) Issues in the haptic display of tool use. In: Proceedings of IEEE/RSJ international conference on intelligent robots systems, pp 140–145

Craig JJ (1989) Introduction to robotics: mechanics and control, 2nd edn. Addison-Wesley, Reading

Duriez C, Dubois F, Kheddar A, Andriot C (2006) Realistic haptic rendering of interacting deformable objects in virtual environments. IEEE Trans Vis Comput Graph 12:36–47

Gregory A, Mascarenhas A, Ehmann S, Lin MC, Manocha D (2000) 6-DOF haptic display of polygonal models. In: Proceedings of IEEE visualization conference, pp 139–146

Hannaford B, Ryu J-H, Kim YS (2002) Stable control of haptics. In: McLaughlin ML, Hespanha JP, Sukhatme GS (eds) Touch in virtual environments, Chap 2. Prentice-Hall, Upper Saddle River, pp 47–70

Ho C, Basdogan C, Srinivasan MA (1999) An efficient haptic rendering technique for displaying 3D polyhedral objects and their surface details in virtual environments. Presence Teleoperators Virtual Environ 8(5):477–491

Hubbard PM (1996) Approximating polyhedra with spheres for time-critical collision detection. ACM Trans Graph 15(3):179–210

Johnson DE, Willemsen P (2004) Accelerated haptic rendering of polygonal models through local descent. In: Proceedings of haptics symposium, pp 18–23

Jones LA, Hunter IW, Irwin RJ (1992) Differential thresholds for limb movement measured using adaptive techniques. Percept Psychophys 52:529–535

Kim YJ, Otaduy MA, Lin MC, Manocha D (2003) Six-degree-of-freedom haptic rendering using incremental and localized computations. Presence 12(3):277–295

Lin MC, Otaduy M (2008) Haptic rendering: foundations, algorithms, and applications. A K Peters, Ltd, USA

Mahvash M, Hayward V (2005) High-fidelity passive force-reflecting virtual environments. IEEE Trans Robot 21(1):38–46

McNeely W, Puterbaugh K, Troy J (1999) Six degree-of-freedom haptic rendering using voxel sampling. In: Proceedings of ACM SIGGRAPH

Miller BE, Colgate JE, Freeman RA (2000) Guaranteed stability of haptic systems with nonlinear virtual environments. IEEE Trans Robot Autom 16(6):712–719

Nocedal J, Wright SJ (2006) Numerical optimization, 2nd edn. Springer, Berlin

Ortega M, Redon S, Coquillart S (2007) A six degree-of-freedom god-object method for haptic display of rigid bodies with surface properties. IEEE Trans Vis Comput Graph 13(3):458–469

Pang X, Tan HZ, Durlach NI (1991) Manual discrimination of force using active finger motion. Percept Psychophys 49(6):531–540

Redon S (2004) Fast continuous collision detection and handling for desktop virtual prototyping. Virtual Reality 8(1):63–70

Ruffaldi E, Morris D, Edmunds T, Barbagli F, Pai D (2006) Standardized evaluation of haptic rendering systems. In: IEEE haptic interfaces for virtual environment and teleoperator systems, VR

Ruspini D, Khatib O (2000) A framework for multi-contact multi-body dynamic simulation and haptic display. In: Proceedings of IEEE/RSJ international conference on intelligent robots systems, pp 1322–1327

Ruspini DC, Kolarov K, Khatib O (1997) The haptic display of complex graphical environments. In: Whitted T (ed) Proceedings of SIGGRAPH 97, computer graphics proceedings, annual conference series. Addison Wesley, Reading, pp 345–352

Salcudean SE, Vlaar TD (1994) On the emulation of stiff walls and static friction with a magnetically levitated input/output device. In: Proceedings of ASME haptic interfaces for virtual environment and teleoperator systems, pp 303–310

Tan HZ, Durlach NI, Beauregard GL, Srinivasan MA (1995) Manual discrimination of compliance using active pinch grasp: the roles of force and work cues. Percept Psychophys 57:495–510

Tang M, Kim YJ, Manocha D (2009) C2A: controlled conservative advancement for continuous collision detection of polygonal models. ICRA, pp 849–854

Wan M, McNeely WA (2003) Quasi-static approximation for 6 degrees-of-freedom haptic rendering. In Proceedings of IEEE visualization conference, pp 257–262

Wang D, Zhang X, Zhang Y, Xiao J (2011) Configuration-based optimization for six degree-of-freedom haptic rendering for fine manipulation. In: Proceedings of IEEE International Conference on Robotics and Automation (ICRA), pp 906–912

Wang D, Zhang Y, Hou J, Wang Y, Lü P, Chen Y, Zhao H (2012a) iDental: a haptic-based dental simulator and its preliminary evaluation. IEEE Trans Haptics 5(4):332–343

Wang D, Liu S, Zhang X, Zhang Y, Xiao J (2012b) Six-degree-of-freedom haptic simulation of organ deformation in dental operations. In: IEEE international conference on robotics and automation, ICRA 2012, St. Paul, 14–18 May 2012, pp 1050–1056

Wang D, Liu S, Xiao J, Hou J, Zhang Y (2012c) Six degree-of-freedom haptic simulation of pathological changes in periodontal operations. In: 2012 IEEE/RSJ international conference on intelligent robots and systems (IROS2012), Vilamoura, Algarve, 7–12 Oct 2012

Wang D, Zhang X, Zhang Y, Xiao J (2013) Configuration-based optimization for six degree-of-freedom haptic rendering for fine manipulation. IEEE Trans Haptics 6(2):167–180

Weller R, Zachmann G (2009) A unified approach for physically-based simulations and haptic rendering. In: Proceedings of the ACM SIGGRAPH symposium on video games. ACM, New York, NY, USA, pp 151–159

Zhang L (2009) Efficient motion planning using generalized penetration depth computation. PhD thesis, Computer Science Department, University of North Carolina at Chapel Hill

Zhang X, Lee M, Kim YJ (2006) Interactive continuous collision detection for non-convex polyhedra. Vis Comput 22(9):749–760

Zhang X, Wang D, Zhang Y, Xiao J (2011) Configuration-based optimization for six degree-of-freedom haptic rendering using sphere-trees. In: Proceedings of the IEEE International Conference on Intelligent Robot and Systems (IROS), pp 2602–2607

Zilles CB, Salisbury JK (1995) A Constraint-based god-object method for haptic display. In: Proceedings of IEEE/RSJ international conference on intelligent robots and systems, Aug 1995

Chapter 3
6-DoF Haptic Simulation of Geometric Fine Features

In this chapter, we introduce haptic rendering for simulation of fine features, including subtle force feeling caused by contacts at sharp geometric features and manipulation of small-sized objects such as dental calculi. We extend the framework of the configuration-based optimization method introduced in Chap. 2 to simulating sharp features using a multi-resolution sphere tree. We classify the fine features in Sect. 3.1 and then introduce the framework and components in the computational pipeline in Sects. 3.2–3.5. Specifically, in Sect. 3.4.2, we propose a continuous collision detection method to detect collisions with objects composed of very thin parts. This method can avoid the pop-through phenomenon during tool manipulation against thin objects (such as small-sized calculus). In Sect. 3.6, we introduce experimental results to validate the approach.

3.1 Classification and Challenges in Simulating Fine Features

Geometric fine features can be roughly classified into small-sized features (e.g., thin-shell, small calculi), sharp features (e.g., edge, corner, cusp, crack), and textures (e.g., wood surface, sand papers).

Fine features are pervasive in many applications involving training for medical and engineering tasks with dexterous manipulation (McNeely et al. 1999; Otaduy and Lin 2006; Wang et al. 2012a, b, c; Chen et al. 2008). For example, in dental surgery simulation, ridges on surfaces of teeth constitute sharp features (Fig. 3.1a). In virtual assembly of an aircraft engine, splined shafts and holes constitute sharp features (Fig. 3.1b). In haptic rendering of those applications, how to simulate the force and torque feelings of sharp features becomes particularly important to the realism of virtual tasks.

Sometimes, manipulation cannot be accomplished without providing the subtle feeling caused by fine features. In the aircraft assembly task, for instance, in order to insert a splined shaft into a splined hole, the sharp edges of the keys on the splined shaft must be rotated to perfectly align with the sharp edges of the keyways on the splined hole or else it will be stopped by the front surface of the hole. Because the

© Springer-Verlag Berlin Heidelberg 2014
D. Wang et al., *Haptic Rendering for Simulation of Fine Manipulation*,
DOI 10.1007/978-3-662-44949-3_3

(a)   **(b)**

(c)   **(d)**

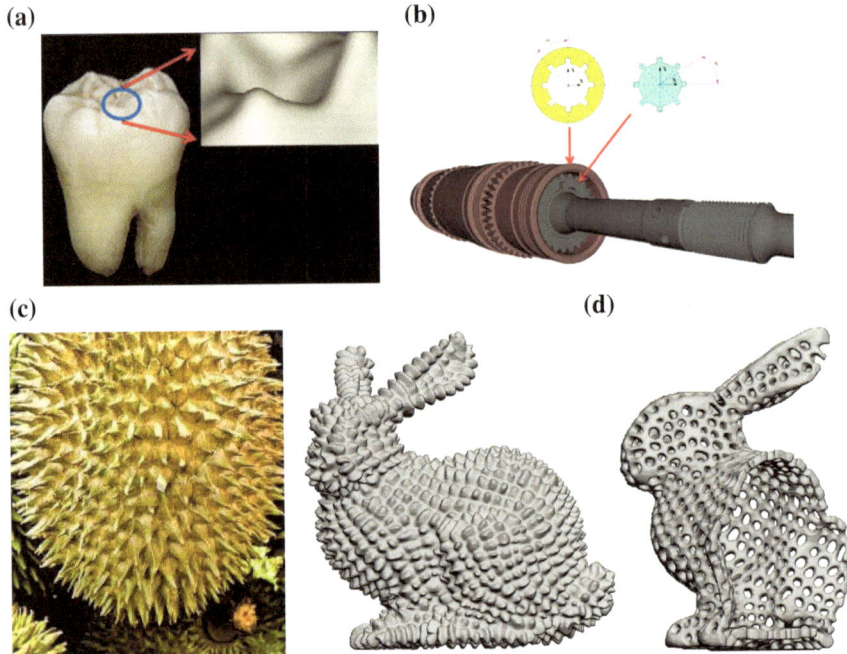

Fig. 3.1 Examples of interactive objects with fine features. **a** Sharp ridges on a tooth. **b** Sharp edges on the splined shaft and hole (*right*). **c** Sharp cusps on a durian. **d** Bumps and small holes

contact region between the splined shaft and the hole is always invisible to the user, the accomplishment of the task will depend on the tactile feeling of the hand (holding the tool) entirely. Therefore, it is necessary to simulate the interaction involving sharp edges/corners between objects with high accuracy and stability.

Similarly, simulation of other fine features is also crucial to the fidelity of haptic simulation. For example, for the small holes with variable diameters on the surface of the bunny in Fig. 3.1d), when using a probe to slide along the surface, the probe can get stuck in some holes if the tip size is smaller than the diameter of the hole.

There are various sharp features in the physical world. In this chapter, we focus on one specific type of sharp features: the sharp edge.

In computer graphics, the types of sharp features are usually divided into four classes: crease, corner, dart, and cusp (Hoppe et al. 1994). Following the definition in computer graphics (Hoppe et al. 1994), we will mainly focus on creases of varying degrees of sharpness for 6-DoF haptic simulation, which could be regarded as the mathematical building block of the other three classes of features and the most common features in haptic tasks.

In computational geometry, a crease or an edge is shaped by two adjacent planes forming an angle, called the dihedral angle β as shown in Fig. 3.2. In haptic fields, some researchers show that it seems impractical to apparently feel the direction variations across a dihedral angle larger than 150° (Hoppe et al. 1994; Shon 2006).

Fig. 3.2 Illustration of a sharp edge for haptic rendering. Copyright © IEEE. All rights reserved. Reprinted, with permission, from Yu et al. (2012)

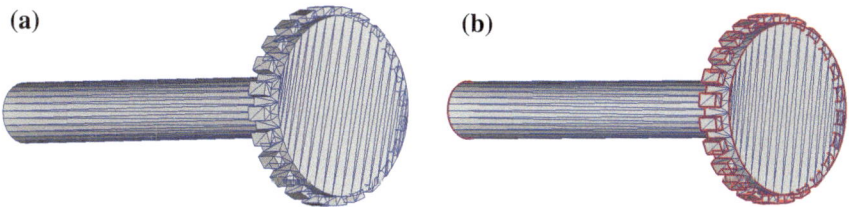

Fig. 3.3 Sharp feature extraction. **a** Mesh model. **b** Sharp edges. Copyright © IEEE. All rights reserved. Reprinted, with permission, from Yu et al. (2012)

Based on this fact, we define a sharp edge as one such that the dihedral angle is within $(0°, 150°]$, as shown in Fig. 3.2.

Because of the monotony of cosine in the angle interval, we just need to calculate the dot product of two normal unit vectors of adjacent planes of each edge. If the result is within $(-1, \sqrt{3}/2)$, it indicates that the current edge is sharp. Take a splined shaft as an example as shown in Fig. 3.3. The original mesh model of the cube includes 931 edges, in which 408 edges are extracted as sharp features with the dihedral angle of 90°, while other edges form the dihedral angle greater than 150°.

One challenge comes from the conflict between model resolution and computation efficiency. Normally, a high-resolution geometric model is needed to represent small-sized geometric features on an object surface, but a large number of geometric primitives in the high-resolution geometric model will increase the computational burden. Another challenge is that tunneling effect may occur with conventional collision detection and collision response methods.

6-DoF haptic simulation of sharp features also poses several challenges that existing haptic rendering methods cannot handle. Penalty-based methods cannot prevent interpenetration such that the virtual tool could traverse through thin objects or features. Virtual coupling (Colgate et al. 1995) is used for maintaining stability but at the expense of a reduced perceptual capability of some geometric details and the associated changes of force, let alone the abrupt changes of the contact force when contacts occur at sharp features. The high computation costs of the previous constraint-based methods (Duriez et al. 2006; Ortega et al. 2007) make those methods unsuitable from dealing with objects with many fine features.

In the constrained-based 6-DoF haptic rendering method introduced in Chap. 2, sphere trees are used to model objects. However, shallow trees with large spheres as

(a)   **(b)**

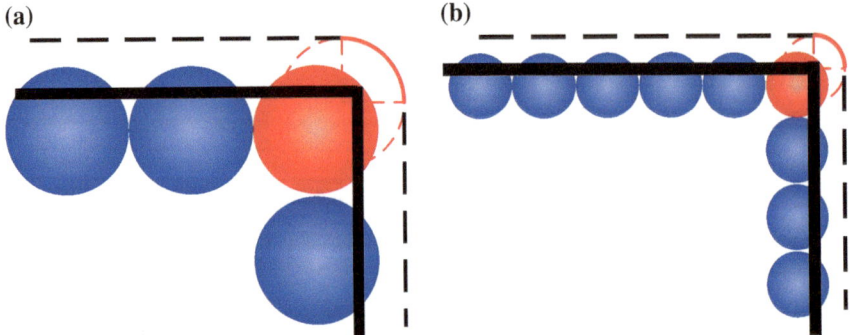

Fig. 3.4 Modeling sharp edges by spheres: Sharp feeling is lost with large spheres (*left*) but better preserved with small spheres (*right*). Copyright © IEEE. All rights reserved. Reprinted, with permission, from Yu et al. (2012)

leaves cannot preserve the feelings of sharp features. For example, as shown in Fig. 3.4, when a probe slides across an edge of a cube, if the sharp edge is represented by large spheres, the sharp feeling is lost.

Thus, a finer sphere-tree model is needed. However, the finer model we need, the more levels of a sphere tree and the larger number of the spheres would be required. For example, to model a rectangular splined shaft, an octree of spheres can be obtained by iteratively dividing all the leaf spheres until each sphere size is below the perception threshold. However, such an octree requires a tree level of 6 (with level 0 being the root node) at least, and the number of spheres in the tree can reach 300,000. In that case, the haptic update rate will be reduced to far smaller than 1 kHz. Although the memory cost could be reduced partly by constructing a non-uniform full octree, there will still be a huge memory waste since a large number of leaf spheres are unnecessarily small for large, smooth regions.

The key questions for simulating sharp features are as follows:

- Can we quantify the influence of sphere size on force feeling of sharp features from the perspective of human perception threshold?
- Can we construct a new sphere-based model to preserve the force feeling produced by the sharp features while maintaining real-time performance?

3.2 Proposed Approach for Simulating Sharp Features

We address those questions in this chapter. We extend the configuration-based optimization method with the sphere-tree model by introducing an adaptive perception-based threshold of sphere size to approximate various sharp features; we also introduce a sphere splitting method to produce the subtle force feeling at sharp features with real-time performance. Furthermore, we add a perception-based experiment to quantitatively analyze the improvement and a haptic-to-vision shape

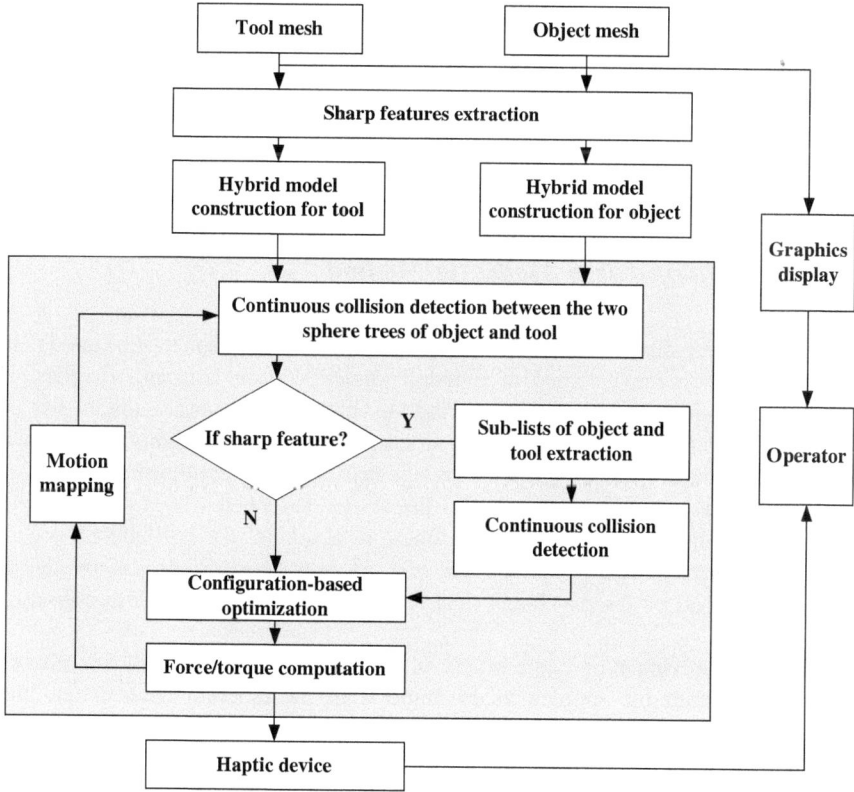

Fig. 3.5 The flowchart of the method. Copyright © IEEE. All rights reserved. Reprinted, with permission, from Yu et al. (2012)

matching task to qualitatively compare the performance between the method and other rendering methods. Figure 3.5 shows the flow chart of the method.

In the off-line precomputation phase, our method first extracts all the sharp features of the object and tool models and then creates a sphere octree for each object or tool model to represent its overall shape without sharp features. For sharp edges, if the radius of a sphere on the leaf node level intersected with the sharp edges is greater than a radius threshold (will be explained in Sect. 3.3.1), the sphere will be split into several smaller spheres stored in a list by a splitting algorithm.

In the haptic loop, the so-called region-sensitive collision detection is performed first, which means if collision occurs in the non-sharp region, only the two sphere octrees of the object and tool will be involved in the collision detection process, or else, the sphere lists will also be checked for collision detection of sharp features. With the novel hybrid sphere tree, the number of spheres for model representation and force computation can be remarkably reduced, while sharp features can also be well reserved in the model to benefit real-time haptic rendering of subtle force sensation of sharp features.

3.3 Multi-resolution Sphere-Tree Model for Sharp Features

The geometric models for the tool and the objects are essential for the simulation method. In this section, we describe in detail the hybrid sphere tree to model both the global shape and local fine features of an object.

3.3.1 Perception-based Modeling Method

Given an object represented by a triangle mesh, a hybrid sphere-tree model is created based on the variance of dihedral angle between adjacent triangles. It consists of a sphere tree for representing the global shape and a linked list of spheres for representing sharp features. In each local area with sharp features, we first identify those spheres with radii greater than a perceptual threshold. Each of those spheres is represented by a linked list of smaller spheres.

Obviously, the size of the smallest sphere in a sphere tree will affect the perception of sharp features. The proper size of a sphere to approximate a sharp edge is determined based on the psychophysical results of human perception threshold of sharp features.

The pilot experiment of Yamashita et al. indicates that perception of smoothness is associated with the supplementary angle α of the dihedral angle β and the maximum distance d between a smooth shape, such as a spherical or cylindrical surface, and its tangent facets, as shown in Fig. 3.6 (Yamashita and Lindeman 2000; Yamashita et al. 2001). In their experiments, a sphere or cylinder is the ideal curved shape to be simulated, and the triangle mesh is the approximate model with adjustable dihedral angles. The goal of their experiment is to find the perception threshold of the dihedral angle so that smoothness of the ideal shape can be simulated using the triangle mesh without perceptible artifacts.

Different from the work in Yamashita and Lindeman (2000), we are facing a reversed situation in that the mesh model with sharp features is the target model, while the sphere-tree model is the approximate model with adjustable radii. We aim

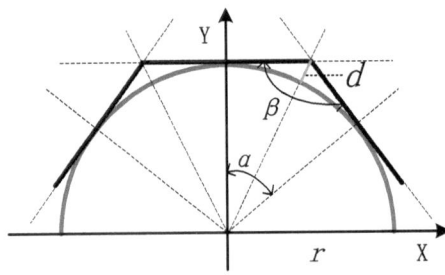

Fig. 3.6 Cross-sectional illustration of a spherical surface and its approximate polygonal model (Yu et al. 2012)

Table 3.1 Some radius threshold samples of sphere

β (degree)	30	60	90	120	150
α (degree)	150	120	90	60	30
r (mm)	1.4	1.8	2.3	3.5	7.4

to find out an acceptable value of a sphere's radius for the sphere-tree model to approximate the target model based on human perception threshold on sharp edges. Therefore, the result of the former experiment can be used as a guideline to determine the sphere size for our method.

The results of the experiments in Yamashita and Lindeman (2000) show that if the force is shaded by interpolating the force signals around the sharp edge, the perception threshold is mainly determined by the maximum distance d. If there is no force shading, the perception threshold depends on the absolute threshold angle α that is proportional to curvature c (which equals $1/r$, and the unit is $1/mm$) as shown below under a certain stiffness of 0.5 N/m (Yamashita and Lindeman 2000):

$$\alpha = 203 \times c + 2.78 \tag{3.1}$$

Based on the limit on the dihedral angle $\beta \in 0°, 150°$ for a sharp edge, we obtain $\alpha \in [30°, 180°)$. With the reciprocal relation between c and the maximum radius r of a sphere as the approximate model, we can determine the size threshold of a sphere for modeling each sharp edge with a dihedral angle within the human perception threshold as the following equation:

$$r \le 203/(180 - \beta - 2.78) \tag{3.2}$$

This is a fundamental criterion for the sphere splitting method below. According to (3.2), some radius threshold r is computed against the dihedral angle β as calculated in Table 3.1.

3.3.2 Hybrid Sphere-Tree Construction

As shown in Fig. 3.7, we add linked lists to some leaf nodes of the sphere-tree model introduced in the Chap. 2 to represent sharp edges. More specifically, from the sphere-tree model representing the overall shape of an object, we identify each leaf sphere that intersects a sharp edge and then perform sphere splitting on the leaf sphere to form a linked list of smaller spheres to represent the sharp edge.

Figure 3.8 shows the results for two examples: a cube and a splined shaft represented in different level of details. Figure 3.9 shows two example objects after adding the linked list to the original sphere-tree model: a cube and a splined hole. As shown in the figure, the sharp edges of both models are better represented by the introduced hybrid sphere-tree modeling method.

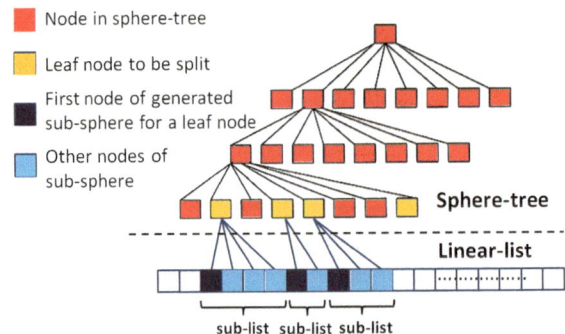

Fig. 3.7 Hybrid sphere tree model consists of a hierarchical sphere tree and a linear list. Copyright © IEEE. All rights reserved. Reprinted, with permission, from Yu et al. (2012)

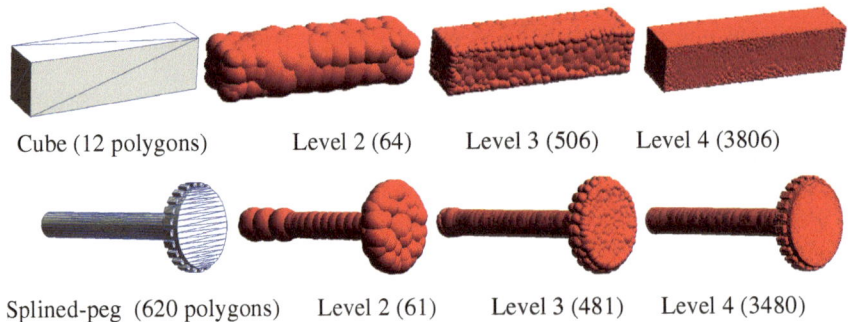

Fig. 3.8 Example models with different levels of details. Copyright © IEEE. All rights reserved. Reprinted, with permission, from Yu et al. (2012)

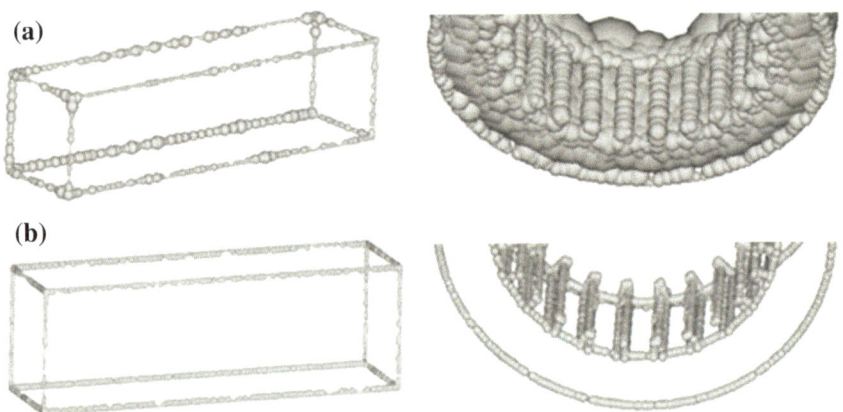

Fig. 3.9 Models for the edges of a cube and a splined hole. **a** The extracted spheres at sharp edges from the 4th level of the sphere octrees. **b** The hybrid models at sharp edges with the method (where r is restricted to 2.3 mm for the dihedral angle of 90°). Copyright © IEEE. All rights reserved. Reprinted, with permission, from Yu et al. (2012)

3.3.3 Sphere List Construction

Algorithm 3.1 outlines the sphere splitting process. To start, the radius threshold r_t of the spheres for each sharp edge is calculated according to Eq. (3.2), which is used to decide the size of the spheres for each edge in the linked list.

ALGORITHM 3.1
SPLIT SPHERES FOR SHARP EDGES

---

input: all the spheres at the leaf level of sphere tree model S , all
the sharp edges E

linked list $S_{\text{list}} \leftarrow \phi$

for each sharp edge $e_i \in E$ of two adjacent facets F_1 , F_2

compute the radius threshold r_t for e_i

for each leaf sphere $s_i \in S$

if (s_i intersects e_i)

if($s_{i \cdot \text{radius}} < r_t$)

then shift s_i to be beneath F_1 and F_2

else $n_s \leftarrow l_c / (r_t \times 2) + 1$

create $numsplit$ of spheres s_{children} for s_i

arrange s_{children} beneath F_1 and F_2 equidistantly along e_i and push

s_{children} to S_{list}

save s_{children} 's head pointer in S_{list} ,n_s and r_t into s_i

move the center of s_i to be the center of the middle sphere of s_{children}

output: S_{list} , updated S

For each leaf sphere modeling a sharp edge, as shown in Fig. 3.10a, if the radius of the sphere is smaller than r_t, splitting is not needed, and the sphere is simply shifted to be under the two adjacent faces of the sharp edge, as shown in Fig. 3.10b.

If the radius of the sphere is greater than r_t, it is split into several smaller spheres of the same size as defined by r_t to form a sphere list. Next is to decide (1) the number n of the generated sub-spheres from splitting a leaf sphere at a sharp edge and (2) how to store and fast access these sub-spheres.

To decide n, we first denote l_c as the chord length of the edge segment cut by the intersection with the leaf sphere before splitting (Fig. 3.10c). The number n is computed by the following model

$$n \leq l_c/(2 \times r_t) + 1 \tag{3.3}$$

Such a choice of n keeps all the sub-spheres connected (i.e., intersected) and evenly arranged to cover the chord length l_c.

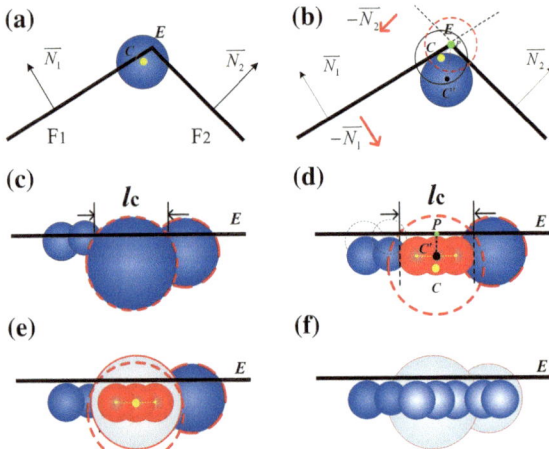

Fig. 3.10 Splitting spheres at a sharp edge. **a** The left view of a sphere and the sharp edge models. **b** A sphere for a sharp edge constrained beneath the two adjacent facets of the edge. **c** Spheres with *red dash line* have radii greater than the threshold. **d** The sub-spheres (*red spheres*) as the result of splitting tightly fit the original mesh model. **e** After splitting, the original sphere as the parent node of sub-spheres is shifted to include the sub-spheres in its center. **f** The finer model of the sharp edge is constructed as the *blue spheres*. Copyright © IEEE. All rights reserved. Reprinted, with permission, from Yu et al. (2012)

To decide how to store and access the sub-spheres, the projection point P of the leaf sphere's center C to the sharp edge is calculated as the first step. Then, the projection point P is moved reversely along the two normals of adjacent facets with the distance r_t to achieve the point C' as the center of all the generated sub-spheres (Fig. 3.10d). In this way, all the generated sub-spheres, regarded as the child nodes of the leaf sphere, could be effectively restricted no larger than the mesh surface. Next, the sub-spheres are pushed into a linked list in sequence; the pointer to the head of list, number of sub-spheres, and radius of each sub-sphere (the same as r_t) are saved back to the leaf sphere for fast access. The last step is to make the center C of the leaf sphere to be at the center C' of the sub-spheres, which makes the leaf sphere covering all the generated sub-spheres, the same way as the parent sphere covers all its child spheres (Fig. 3.10e). After the above procedures for each leaf sphere modeling all the sharp edges, a finer hybrid sphere-tree model is successfully established as shown in Fig. 3.10f.

3.4 Collision Detection for Objects with Fine Features

In this section, we introduce an efficient collision detection method based on the hybrid sphere-tree model introduced above. Furthermore, we introduce a sphere-tree-based continuous collision detection (SCCD) method to simulate manipulation against small-sized features to avoid tunneling effects.

3.4.1 Hybrid Sphere-Tree-Based Collision Detection

There are two steps for collision detection between two objects with sharp edges. The first step is for tree–tree detection, as described in Chap. 2, and the second is for list–tree or list–list detection. The output of the collision detection is a set of intersected pairs of leaf spheres for the global shape and/or sub-spheres for sharp edges.

In each collision detection process, if an intersected leaf sphere is found to involve a sharp edge by pointing to a linked list of sub-spheres from the step of tree–tree detection, the step of list–tree or list–list detection begins.

The sub-spheres of the intersected leaf sphere are then extracted from the linked list to check against either the tree or another linked list of the other object for collision.

A list–list has a worst-case time complexity of $O(n^2)$. Since n is relatively small, in tens, collision detection is efficient enough to maintain the update rate required by haptic rendering.

3.4.2 Sphere-Tree-Based Continuous Collision Detection

Let s_T represents a tool sphere at any level and s_O represents an object sphere at the same level. If no intersection is detected between them at discrete time instants $t - 1$ and t, it is still possible that a collision happens between the time instants $t - 1$ and t if the object is very thin (see Fig. 3.11). This phenomenon is called tunneling effect (Lin and Otuday 2008).

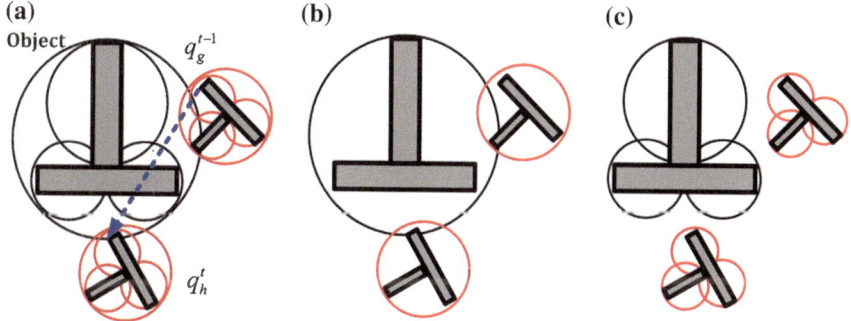

Fig. 3.11 Tunneling effect: Even though no collision is detected between the two T-shape objects, at time instants $t - 1$ and t, **a** shows that a collision exists between $t - 1$ and t. Copyright © IEEE. All rights reserved. Reprinted, with permission, from Wang et al. (2014)

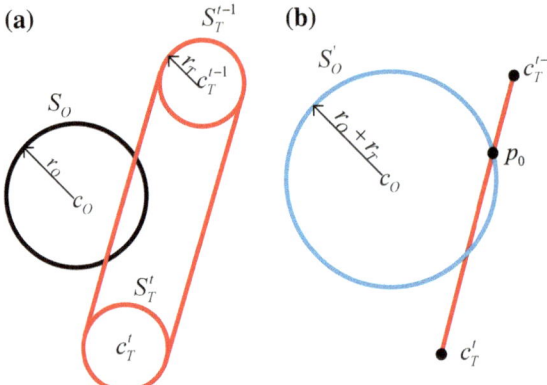

Fig. 3.12 Collision check between the moving tool sphere s_T and the fixed object sphere s_O shown in **a** becomes an intersection query between the line segment and the enlarged sphere in the tool's configuration space shown in **b**. Copyright © IEEE. All rights reserved. Reprinted, with permission, from Wang et al. (2014)

To prevent the tunneling effect, we introduce a continuous collision checking method between s_T and s_O as follows. First conduct a linear interpolation between s_T^{t-1} and s_T^t, as shown in Fig. 3.12a. In the configuration space of s_T, the sphere shrinks to a point, and thus, its swept volume between s_T^{t-1} and s_T^t becomes a line segment, whereas the object sphere s_O grows by the tool's radius, as shown in Fig. 3.12b. Thus, the collision query between the moving tool sphere and the fixed object sphere can be done by simply checking the intersection between the line segment and the enlarged object sphere. We call this intersection check between a line segment and a sphere *lineSeg-sphere check*. If we define a relative time index $t_{OC} \in [0, 1]$ in one sampling period, then the relative time when first contact occurs can be computed as:

$$t_{oc} = \frac{\overline{c_T^{t-1}p_0}}{\overline{c_T^{t-1}c_T^t}} \tag{3.4}$$

We use the *lineSeg-sphere check* described above instead of a simple sphere–sphere check at every step and level of collision check between the sphere tree of the tool and that of the object, which we call SCCD. The SCCD algorithm first carries out the lineSeg-sphere check for the root bounding spheres of the tool and the object. If there is an intersection, it goes down to their children nodes and query collisions using their respective spheres, and so on, until leaf nodes are reached.

For collision query at the level of leaf nodes, t_{oc} is recorded for each intersection. The minimum value of t_{oc} is set as the first collision time. The configuration of the haptic tool at t_{oc} is the computed result of the SCCD.

3.5 Fast Collision Response with PQP

The collision response approach introduced in Chap. 2 is based on optimization
using the active set method, which is generally most effective for small- to medium-
scale quadratic programming (QP) problems. The optimal solution can be obtained
in a finite number of iterations, but a good start can shorten this process greatly. For
a common model with 4,681 spheres (an octree with 4 levels), if more than 100
pairs of spheres intersect, the update rate of haptic simulation will decrease to be far
below 1 kHz, which is also not appropriate for simulating large-area/volume contacts in real time. The time cost of optimization becomes the bottleneck to real-time
haptic feedback.

In this section, we introduce a parallel quadratic programming (PQP) method
(Brand et al. 2011) to accelerate the optimization process especially for cases with
large-area contacts.

Figure 3.13 illustrates the general flowchart of the configuration-based optimization approach for 6-DoF haptic rendering with PQP. Most QP problem solver
requires long computation time at each iteration. Although many reports in the
literature address speed-up methods (Milman and Davison 2008; Richter et al. 2009),
they are heuristics without any guarantees on convergence to the global minimum.
The PQP method is simple and results in fast convergence—two matrix-vector

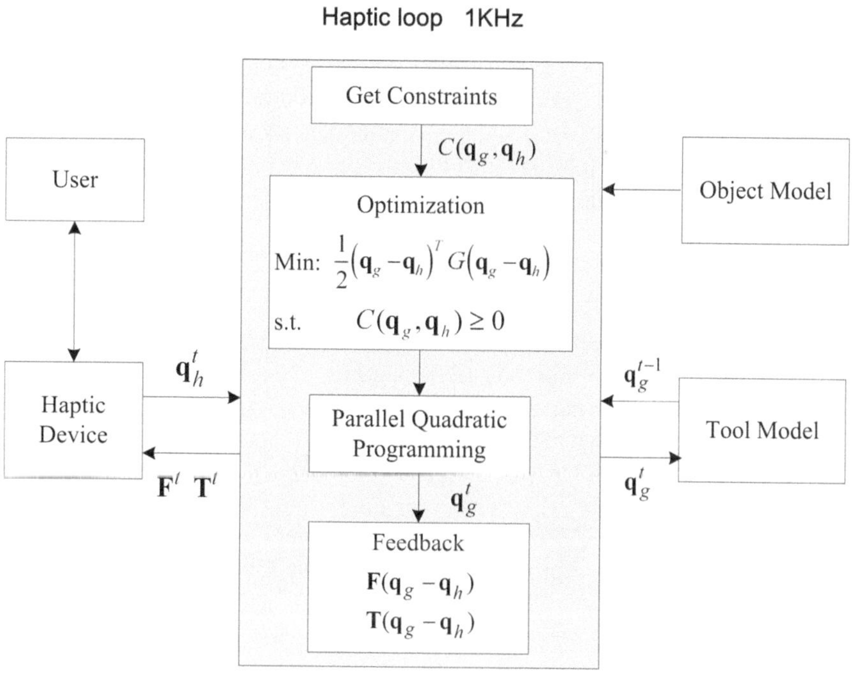

Fig. 3.13 Framework for the configuration-based optimization with parallel quadratic
programming

products and a scalar divide, which offers considerable speed advantages than the
active set method. Besides, PQP does not need to be transformed back and forth
between primal and dual space, which makes PQP particularly efficient. Before we
proceed to solve the QP problem, we first demonstrate how to satisfy the assumptions
in PQP as below:

- The matrix $\mathbf{G}$ should be positive definite in Eq. (2.3).
- The primal QP problem in Eq. (2.3) should be feasible.

In the optimization model, $\mathbf{G}$ is a diagonal matrix formed from the stiffness of
three translational springs and three torsional springs, which meets the first condition. The optimization objective satisfies the principle of minimum total potential
energy, i.e., a structure or an object shall move to a position that minimizes the total
potential energy, which means that there exists a solution to the QP problem.

We now use the PQP method to convert the primal form in Eq. (2.3) into its dual
form given below:

$$\begin{cases} \text{Min: } F(\mathbf{y}) = \frac{1}{2}\mathbf{y}^T\mathbf{Q}\mathbf{y} - \mathbf{y}^T\mathbf{d} \\ \text{Subject to: } \mathbf{y} \geq \mathbf{0} \end{cases} \tag{3.5}$$

where $y \in \mathbb{R}^N$ is the dual variable, N refers to the number of intersected sphere
pairs, and $\mathbf{Q} \in \mathbb{R}^{\mathbf{N} \times \mathbf{N}}$ is also positive semi-definite with

$$\mathbf{Q} = \mathbf{J} \cdot \mathbf{G}^{-1} \cdot \mathbf{J}^\mathbf{T} \tag{3.6}$$

Next, with any initial guess $\mathbf{y}^0 > \mathbf{0}$ (here, we choose $\mathbf{y}^0 = [0.001, 0.001, \dots$
$0.001]^T$), the iterations are performed by the multiplicative update rule (3.7) to solve
the dual problem of (2.3) to a specified tolerance of $\Delta F(\mathbf{y})$ (we assigned 0.01 to it).
The division and the max(a, b) operations are carried out in an element-wise
manner.

$$\begin{cases} \mathbf{y}_{t+1} = \mathbf{y}_t \left[\frac{\mathbf{d}_t^+ + \mathbf{Q}_t^- \mathbf{y}_t}{\mathbf{d}_t^- + \mathbf{Q}_t^+ \mathbf{y}_t} \right] \\ \mathbf{Q}^+ = \max(\mathbf{Q}, 0) + \text{diag}(r) \\ \mathbf{Q}^- = \max(-\mathbf{Q}, 0) + \text{diag}(r) \\ d^+ = \max(\mathbf{d}, 0) \\ d^- = \max(-\mathbf{d}, 0) \\ \text{diag}(r) = \{r_1, \dots r_N\}, r_i = \max(\mathbf{Q}_{ii}, \sum_j \mathbf{Q}_{ij}^-) \end{cases} \tag{3.7}$$

Then, the optimum $(\Delta \mathbf{q}^t)^*$ of the primal problem in (2.3) can be recovered from
the optimum y^* of the dual problem in (3.5) using the following equation:

$$(\Delta \mathbf{q}^t)^* = \mathbf{G}^{-1} \cdot \mathbf{J}^T \mathbf{y}^* \tag{3.8}$$

Finally, we can obtain the desired q_g^t at current time step t:

$$\mathbf{q}_g^t = \mathbf{q}_h^t + (\Delta \mathbf{q}^t)^*$$ (3.9)

3.6 Performance Analysis on Simulating Sharp Features

A Phantom Premium 3.0/6-DoF is utilized as the haptic device to provide six-dimensional forces and torques. Figure 3.14 shows the experimental setup. The specifications of the computer are Intel(R) Core(TM) 2,2.20 GHz, 2 GB memory, and X1550 series radeon graphical card. We have conducted two perception-based experiments and two objective evaluation experiments to validate the introduced method.

Perception-based experiments and a haptic-to-vision shape matching task are used to compare the performance between the method and other rendering methods. The experiment results from a cylinder–cube interaction and a spline-shaped peg–hole interaction validate that the method can simulate subtle force direction changes when an object slides across sharp edges. The comparison results show that the introduced method is effective in simulating sharp features both in terms of measured force signals and human subjective evaluation. Non-penetration among objects is maintained for multi-region contact scenarios. The haptic rendering rate is about 1 kHz, and the interaction is stable.

3.6.1 Perception Experiments: Comparative Study

Perception-based experiments for feeling sharp edges are designed to compare three target methods of haptic rendering:

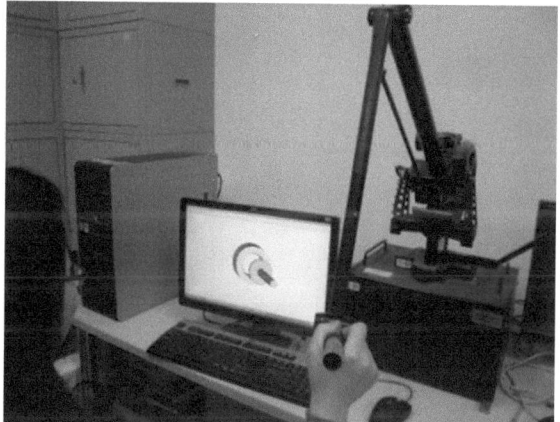

Fig. 3.14 Experimental setup for haptic simulation of fine features. Copyright © IEEE. All rights reserved. Reprinted, with permission, from Yu et al. (2012)

- Ref. 1: the VPS (penalty-based voxmap and point shell method with virtual coupling) (Barbič and James 2008)
- Ref. 2: the constraint-based method with sphere-tree models (in Chap. 2)
- Ref. 3: the introduced method based on hybrid sphere-tree models

We introduce the concept of quasi-curved radius (QCR) as the quantified evaluation metric. A cube modeled by a triangle mesh is employed as the ground truth for comparing perception of sharp edges. As shown in Fig. 3.16, the degree of sharpness is modeled by a curved chamfer of radius r_{cc} varied from 5.0 mm to zero. Human subjects are required to compare the force feeling between the sphere-tree model and the mesh model, and QCR denotes the matched radius of the chamfer perceived by the human subjects.

For each target method, three repeated sessions were performed by each participant. In each session, a number of trials were performed to find the QCR. In each trial, a subject was asked to slide the sphere tool across the edge of two candidate cubes along the red dashed line as shown in Fig. 3.15. The two candidate cubes looked the same, while one was modeled by a sphere tree and the other was modeled by a reference mesh. The participant was asked to compare the force feeling and to respond "yes" if the two shapes were felt the same, otherwise to respond "no." After the response, a new reference mesh model was generated to produce a sequence of reference mesh model in which chamfer radius progressively decreased from 5.0 to 0 mm in value. At the end of the session, a converged chamfer radius was determined, which was defined as the QCR of the target method.

Ten right-handed participants (6 male and 4 female, the average age being 28 years old) took part in the perception experiments. All of them were familiar with the Phantom device. For each participant, it typically took 10 min to finish the three sessions for a target method.

In the above process, the staircase method (Gescheider 1997) was used to adjust the chamfer radius adaptively. When a subject first responded yes in a descending sequence of r_{cc} by 0.2 mm, the program would start an ascending sequence which ended when the subjects first said no, and the sequence would be reversed by 0.2 mm again. This procedure continued until responses of three transitions were recorded. After that, the step was decreased to 0.1 mm and the process continued

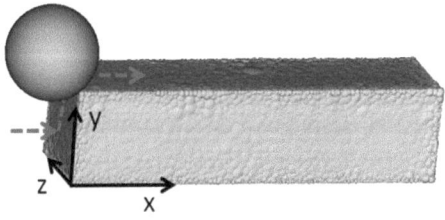

Fig. 3.15 Objects used in the perceptual experiments for sharp edges. Copyright © IEEE. All rights reserved. Reprinted, with permission, from Yu et al. (2012)

Table 3.2 Models in three haptic rendering methods

Methods	Model type	Number of primitives		
Ref. 1	Sample points	$80 \times 80 \times 80$		
Ref. 2	Sphere tree	4,681		
Ref. 3	Sphere tree with linked lists	Sphere tree	4,681	
		Linear list	827	

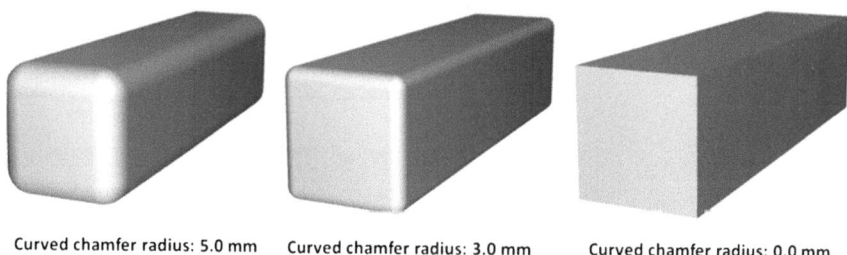

Curved chamfer radius: 5.0 mm   Curved chamfer radius: 3.0 mm   Curved chamfer radius: 0.0 mm

Fig. 3.16 Three examples of cubes with curved chamfer of varied sizes for comparison. Copyright © IEEE. All rights reserved. Reprinted, with permission, from Yu et al. (2012)

until another four times of response transitions were recorded. Finally, the threshold r_{cc} was taken as the average of the transition points. The r_{cc} at this time was recorded as the QCR of the target method.

The same object and tool were used for testing each rendering method: a cube of 120 mm × 30 mm × 30 mm as the interacted object, and a sphere with radius 30 mm as the moving tool (Fig. 3.15). The three kinds of geometric models used in the three haptic rendering methods are shown in Table 3.2. As a reference for perception comparison, a triangle mesh model and a point tool were employed. Different from the used cube model above, the triangle mesh is modeled as a cube with a curved chamfer, in which the cube shares the same size while the curved chamfer of varied sizes, as shown in Fig. 3.16. All the three haptic rendering methods were tested with the same virtual stiffness of the force computation model.

The mean and standard deviation of QCR for the three target methods are presented in Fig. 3.17. One-way ANOVA was performed to compare the three haptic rendering methods, which showed statistically significant differences. Between Ref. 3 and Ref. 2, the statistical analysis result is $t(9) = 8.81$, $p < 0.05$. Between Ref. 1 and Ref. 2, the statistical analysis result is $t(9) = 6.75$, $p < 0.05$. Between Ref. 3 and Ref. 1, the statistical analysis result is $t(9) = 11.07$, $p < 0.05$.

The QCR value of Ref. 1 is the greatest among the three methods. It implies that a square, sharp edge is felt like a spherical surface with radius 3.8 mm, which is far from the real sharp feeling. The cause maybe that virtual coupling filtered the subtle force feedback of sharp features. Ref. 2 is better than Ref. 1, but it is still too smooth to simulate the sharp edge. The QCR value of our introduced method Ref. 3 with the sphere tree and the linked list model is the smallest, and the value is also

Fig. 3.17 Percepted radius of a right-angled edge for each of the three haptic rendering methods. Copyright © IEEE. All rights reserved. Reprinted, with permission, from Yu et al. (2012)

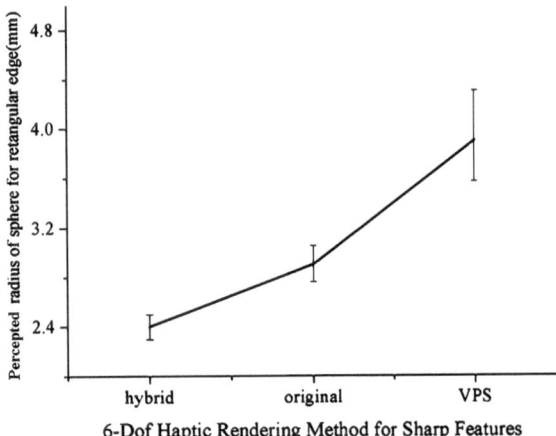

almost the same as the radius limit computed by Eq. (3.2). The results indicate that the proposed method could provide more realistic haptic feeling of sharp edges than the other two methods.

3.6.2 Perception Experiments: Shape Matching

In this experiment, subjects are instructed to assemble an L-shaped peg with a square section into a square hole by haptic perception only.

Four square holes of the same size are visually presented to the participants as shown in Fig. 3.18. Meanwhile, four corresponding different haptic holes are invisibly designed in the experiment, including a square hole, a square hole with a curved chamfer shape, a cylinder hole, and a wedge-shaped hole as shown in Table 3.3. Taking into account the errors between the sphere model and the ideal model, the sizes of the four haptic holes are designed such that the cubic peg (20 mm × 20 mm × 20 mm) could just fit into each one. The four visible square holes have the same size as the maximum size of the four haptic holes, which provide the rough scope for exploring each haptic hole.

When a user moves the handle of a 6-DoF haptic device to explore the virtual surface of the four holes, he/she feels the contact forces and torsional torques while exploring the subtle geometric features of the holes (such as sharp edges, corners, and curvatures) to recognize the right shape. Two methods (Ref. 1 and Ref. 3) were employed to compare the accuracy of shape matching.

Twelve right-handed participants of age 23–30 participated in the experiments. All the participants had previous experience with the haptic device. First, each participant held the peg to aim at each visible square hole and then insert each hole by moving or rotating the peg. They were asked to concentrate on perceiving the subtle feel changes to find the match to one of the haptic holes cued from the graph

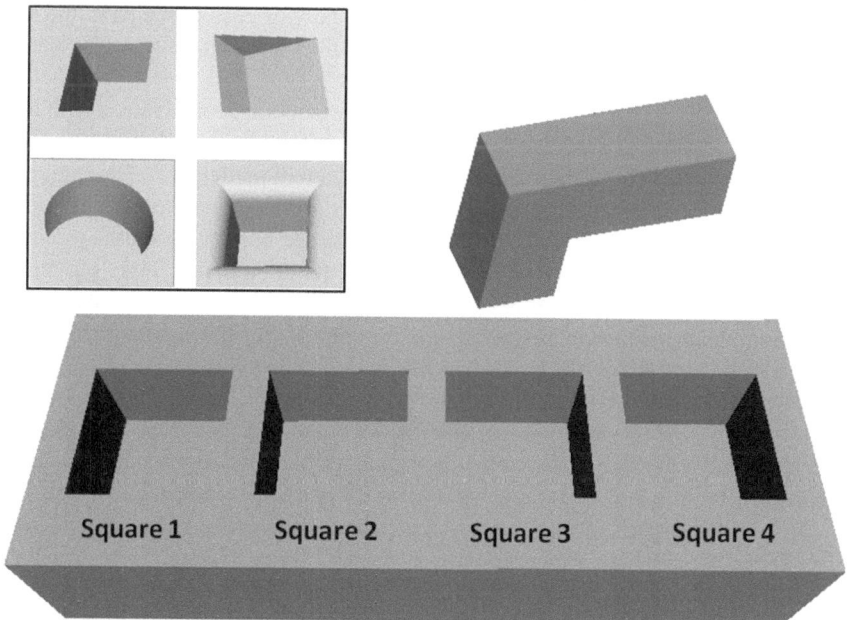

Fig. 3.18 Four visible square holes to explore, which disguise the four haptic holes shown in the upper-left corner. Copyright © IEEE. All rights reserved. Reprinted, with permission, from Yu et al. (2012)

Table 3.3 Models in three haptic rendering methods

Haptic holes	Shape description	Size (mm)
Square1	Square	20.5 × 20.5 × 20.5
Square2	Square with a curved Chamfer	20.5 × 20.5 × 20.5 radius of curved chamfer: 3.5
Square3	Cylinder	Radius: 14.14
		Height: 20.5
Square4	Wedge	Edge length: 20.5
		Height: 20.5

at the top-left corner of the screen. Participants were required to report the closest match based on the haptic feel. If they were not fully certain about the choice, they could provide their best guess. After shape matching, they assembled the cube peg into the square hole they believe to be the right one. For each participant, the test typically lasted for 15 min, which consisted of 6 trials including 3 trials for Ref. 1 and 3 trials for Ref. 3.

Tables 3.4 and 3.5 show the results of all the participants' performance with two methods, respectively, in the matching task. Each table presents the percentage of matching a visible hole to a haptic hole averaged over all participants. For example, 0.0 % means no participant matched the shape they felt in the row to the real haptic shape in the column. The highlighted cells indicate the correct matches.

One-way ANOVA was performed to compare the three haptic rendering methods. For the square, the difference is $t(11) = 2.39$, $p < 0.05$, which showed statistically significant differences between the two methods. For the square with curved chamfer, the difference is $t(11) = 1.96$, $p > 0.05$). For the cylinder, the difference is $t(11) = 1.25$, $p > 0.05$. For the wedge, the difference is $t(11) = 1.36$, $p > 0.05$.

Ref. 3 shows better performance than the VPS method for all the four shapes, especially in distinguishing between the square with sharp edges and the square with curved chamfers. For example, the rates 41.7 and 38.9 % in cell (1, 1) and cell (2, 1), respectively, in Table 3.4 show that a user can hardly make a distinction between the square haptic hole and the one with curved chamfer with the VPS method. The main reason may be that the radius of the curved chamfer (3.5 mm) is below the QCR (3.8 mm) for sharp edge with a right angle in Ref. 1. On the other hand, our introduced method (Ref. 3) can provide better sharp feeling for a user to distinguish the two cases, as evident from the matching rate of 75 % in cell (1, 1) and the rate of 16.7 % in cell (2, 1) of Table 3.5.

Table 3.4 Correct rate of shape matching with Ref. 1 (VPS method)

Shapes of haptic hole	Perceived shape (%, from all 12 participants)			
	Square (%)	Square with curved chamfer (%)	Cylinder (%)	Wedge
Square	41.7	36.1	8.3	13.9
Square with curved chamfer	38.9	44.4	5.6	11.1
Cylinder	8.3	11.1	77.8	2.8
Wedge	11.1	8.4	8.3	72.2

Table 3.5 Correct rate of shape matching with Ref. 3 (the hybrid model method)

Shapes of haptic hole	Perceived shape (%, from all 12 participants)			
	Square (%)	Square with curved chamfer (%)	Cylinder (%)	Wedge (%)
Square	75	5.6	8.3	11.1
Square with curved chamfer	16.7	72.2	2.8	8.3
Cylinder	5.6	8.3	86.1	0.0
Wedge	11.1	5.6	2.8	80.5

Subjects often correctly identify the haptic cylinder hole, see cell (3, 3) in both Tables 3.4 and 3.5, by an axial rotation of the peg after inserting the hole because there is little torque feedback in the procedure. The results imply that torques can also be very helpful in identifying shapes. The wedge-shaped hole is also frequently matched correctly as shown in cell (4, 4) by the exclusive perception of an oblique compliance with its surface.

3.6.3 Experiments for Objective Evaluation

In this section, two example experiments of a cylinder–cube interaction and a spline-shaped peg–hole interaction are performed. The purpose is to validate that the method can simulate subtle force direction changes when an object slides across sharp edges.

3.6.3.1 Example 1: Cylinder–Cube Interaction

In this experiment, we used a 120 mm × 30 mm × 30 mm cube as the static object, and a cylinder with a radius of 4 mm and height of 40 mm as the moving tool. The three reference methods were employed to simulate the cylindrical tool sliding along an edge of the cube, in which the surface of the cylinder keeps the contact, and the center of the tool is denoted as the grasp point of the device. Applying Ref. 1 to make the tool interact with the cube, the configurations of the tool during interaction were recorded as the inputs for the other two rendering methods, so that the tool could maintain the same configurations in the three methods for comparison. The force and torque curves are shown in Figs. 3.19 and 3.20 in which the coordinate system is set the same as the one in Fig. 3.21.

Based on the contact status, we divide the whole process into five stages as below. At stage 1, the cylinder tool begins to contact the $y+$ facet of the cube and slides to the edge vertically on the surface. Therefore, only the force along y axis comes out. Then, the tool slides across the sharp edge from $y+$ facet to $x-$ facet when the grasp point shifts out of the edge at stage 2. As a result, the force along y axis starts to decrease while force along x axis increases. Meanwhile, the torque along z axis grows by the rotation on the y axis (the edge). When the tool explore over the edge, it goes on moving on the $x-$ facet so that only force along x axis is exerted as shown in stage 3.

Next, the tool runs the track reversely back to the $y+$ facet (i.e., the leftmost facet) by crossing the edge again and finally leaving the facet shown in stage 4 and stage 5. We mainly focus on the performance when the tool slides across the sharp edge. Note that both the force and torque curves in the method of Ref. 3 change more rapidly than that in the other methods at stage 2 and stage 4 in Figs. 3.19 and 3.20.

By comparing the gradients of the force curves (of Phase 2) along the axis x, we can clearly observe that the gradient of Ref. 3 is much greater than that of the

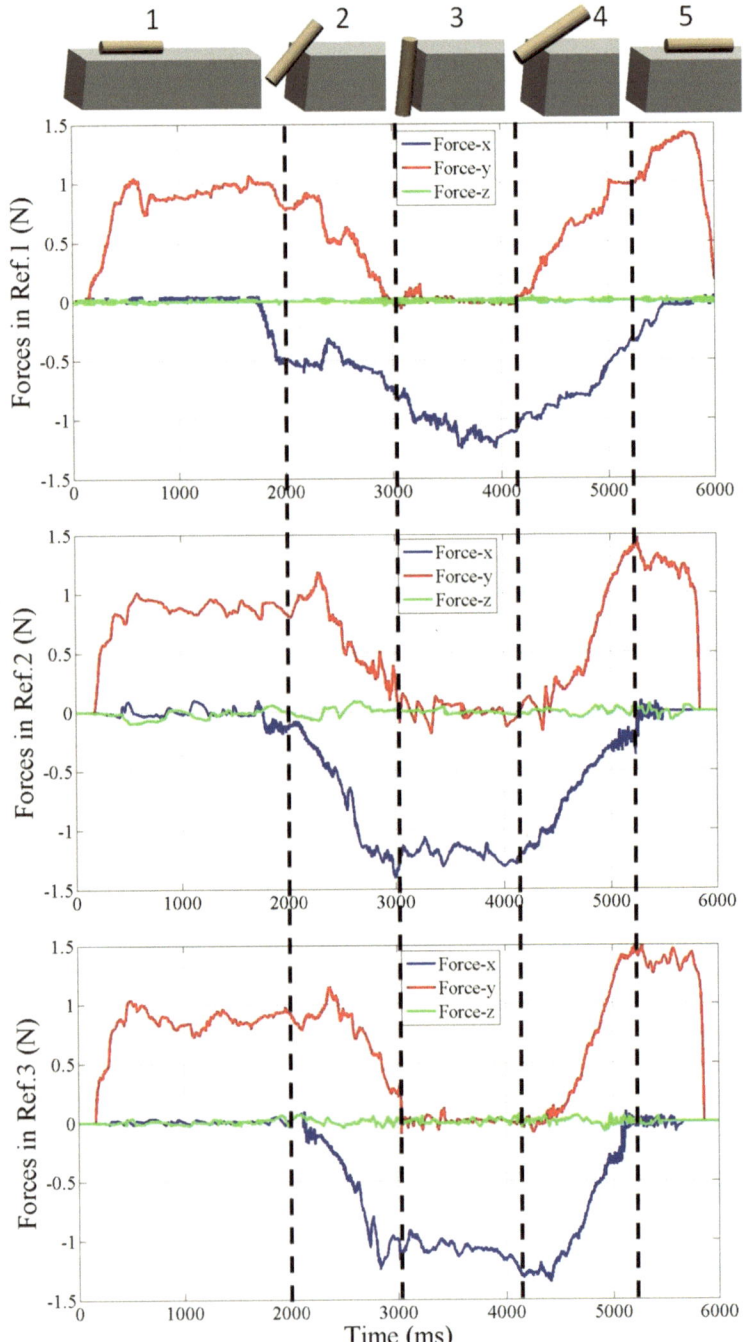

Fig. 3.19 Comparing contact forces produced by the three rendering methods. Copyright © IEEE. All rights reserved. Reprinted, with permission, from Yu et al. (2012)

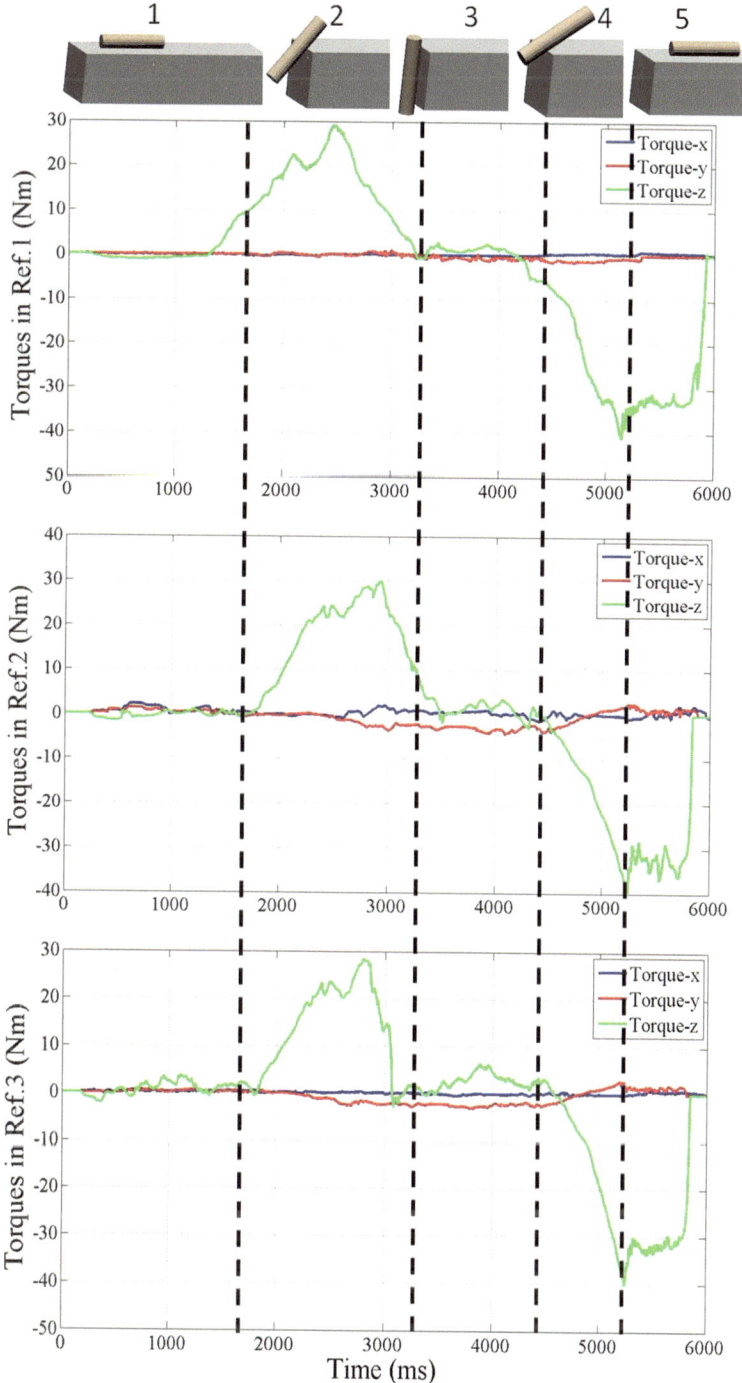

Fig. 3.20 Torques in the three rendering methods for interaction between cylinder and cube. Copyright © IEEE. All rights reserved. Reprinted, with permission, from Yu et al. (2012)

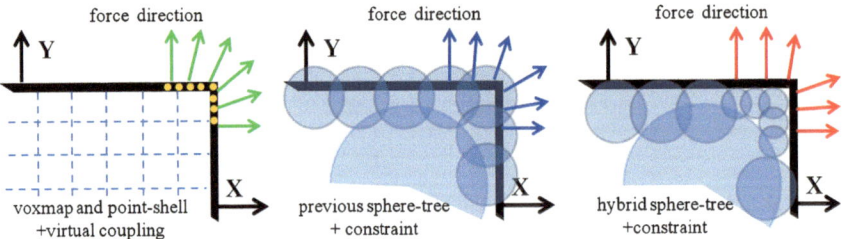

Fig. 3.21 Comparisons of directions of resultant forces from interactions produced by the three rendering methods. Force along the z axis is ignored because its value is too small. Copyright © IEEE. All rights reserved. Reprinted, with permission, from Yu et al. (2012)

Ref. 1, which implies that the directions of resultant forces vary dramatically at the sharp edge (as in real cases) with our introduced method, while the virtual coupling method used in Ref. 1 filtered the force signals against the sharp edge. Similarly, for the torque curves along the axis z, the obvious differences exist between the gradient of Ref. 3 and that of the Ref. 1.

Since the resultant force or torque is what the subjects finally feel, we also show the directions of resultant forces obtained by Ref. 1, Ref. 2, and Ref. 3, respectively, at the same sampling point in Fig. 3.21. The directions of resultant forces vary most dramatically at the sharp edge with Ref. 3. Note that sharp features are characterized by the degrees of variation of normal directions (Boulch and Marlet 2012); hence, the results indicate that our method could reproduce a stronger feeling of sharpness than the other two methods.

3.6.3.2 Example 2: Splined Peg–Hole Interaction

The purpose of this experiment is to validate whether the proposed approach can simulate the insertion process. One user manipulated the haptic device to perform the peg-in-hole insertion task.

To begin with, the hybrid sphere models of splined shaft and hole are shown as Fig. 3.22. The major diameter of the splined hole is 62.5 mm, and the pressure angle is 15°. The splined hole model consists of 8,861 spheres including 4,681 spheres for the sphere tree and 4,180 spheres for the linear list, and the two numbers for the splined shaft are 4,681 and 4,446. Under the number of spheres for interaction, results show that the update rate of the haptic loop can also maintain 1 kHz.

The force curve is shown in Fig. 3.23, which goes through the states of being separate, rotating, and sliding. At the initial stage of the virtual assembly task, the peg and hole are always separate with no force output.

When the user started moving the peg into the hole, the peg stopped in front of the hole's surface due to collision with the sharp keyways. Then, the user rotated the splined peg in order to align it perfectly with the hole based on subtle haptic feeling. The user felt the sharp decrease of the axial force when the splined peg was

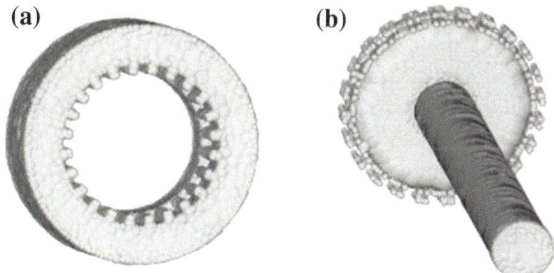

Fig. 3.22 Hybrid sphere model for splined shaft and hole **a** 8,861 spheres **b** 9,127 spheres. Copyright © IEEE. All rights reserved. Reprinted, with permission, from Yu et al. (2012)

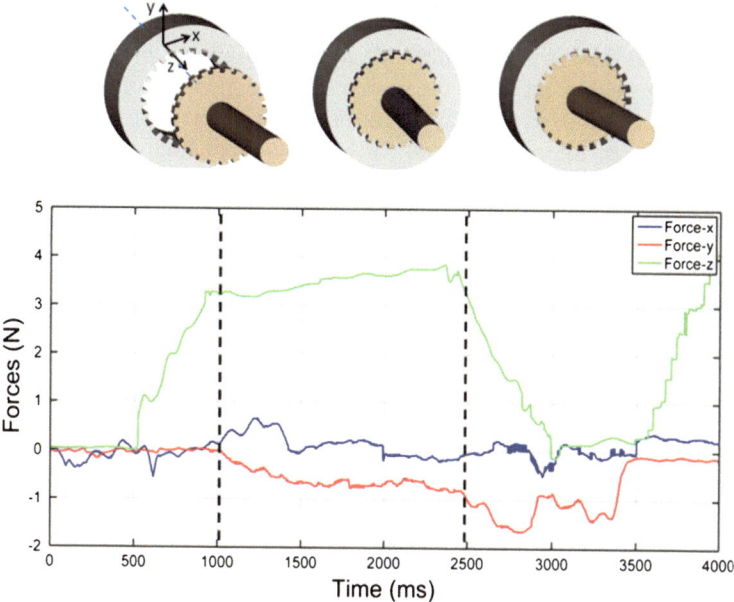

Fig. 3.23 Force signals during three interactive states (separate, rotating, and inserting). Copyright © IEEE. All rights reserved. Reprinted, with permission, from Yu et al. (2012)

rotated to a certain angle, where the sharp keys of the splined peg and keyways of the hole were matched. At this moment, the user pushed the peg into the hole. From Fig. 3.23, we can see that in the rotating state, the Z-directional axial force is the major one.

The user will feel the axial force decrease remarkably when he rotates the splined shaft to some angle, which means the sharp keys of the splined shaft and keyways of hole are totally matched. At this moment, the peg begins to enter into the hole.

In the insertion process, the peg is constrained by the sharp keyways; therefore, it can only move along the axial direction and a little departure from it will make forces along the x and y axes change more frequently, as shown in Fig. 3.23. When the axial force increases again, the peg touches the bottom of the hole which indicates the task is finished successfully. In the whole process, no interpenetration occurs between the graphical peg and hole even at the sharp keys. A stable feedback is also demonstrated by the force curves during interaction.

After the successful insertion of the splined shaft into the hole, the user reported that he could feel the force and torque change when the contact state switched.

3.6.4 Performance of the PQP-based Accelerating Method

In this experiment, we use the Stanford bunny to compare the efficiency of optimization by the *active-set*-based method and the PQP-based method. The golden bunny and silver bunny represent the graphic tool and the fixed object. The original triangle mesh is employed for graphics display, and its sphere-tree model (level 4 with 4,681 spheres for each) is used for haptic computation, including collision detection, constraint modeling, and configuration optimization.

The *active-set*-based method was first applied. In each time step of haptic interaction, the number of intersected pairs of spheres and the time costs of both the collision detection and optimization were saved. Also, the six-dimensional configurations of the haptic tool were recorded so that they would be reused as the input motion later for testing the PQP-based method. In this way, the inputs for the two optimization methods were the same in order to compare the performances of the two methods.

The results are shown in Figs. 3.24 and 3.25. In a shallow contact such as sliding one bunny on the surface of another in this test (with less than 50 intersected sphere pairs), the total time costs of the *active set* and the PQP are less than 1 ms each, i.e., both methods can meet the stringent requirement of 1 kHz update rate of the haptic loop. However, if large area contacts occur (with more than 100 intersected sphere pairs), the total time cost of the *active set* method exceeds 1 ms. Because of the same sphere-tree models and the collision detection algorithm, the two methods took the same time for collision detection. However, the time of optimization with the PQP method occupies only one-third of that with the *active set* method, even less than the time for collision detection. One of the most important reasons for the efficiency with the PQP method comes from the simplicity in the form of the dual problem for the optimization solver, such as avoiding multiple matrix–matrix arithmetic in the *active set*. Although the total time speed-up is not as significant as that for the optimization part alone, the PQP-based rendering method is still twice as fast as the *active-set*-based rendering method.

The average and maximal iterations for solving Eq. (3.7) is six and ten to achieve the optimal solution in the PQP-based optimization. The total time cost depends more on the constraints matrix $\mathbf{J}$ rather than the initial variables.

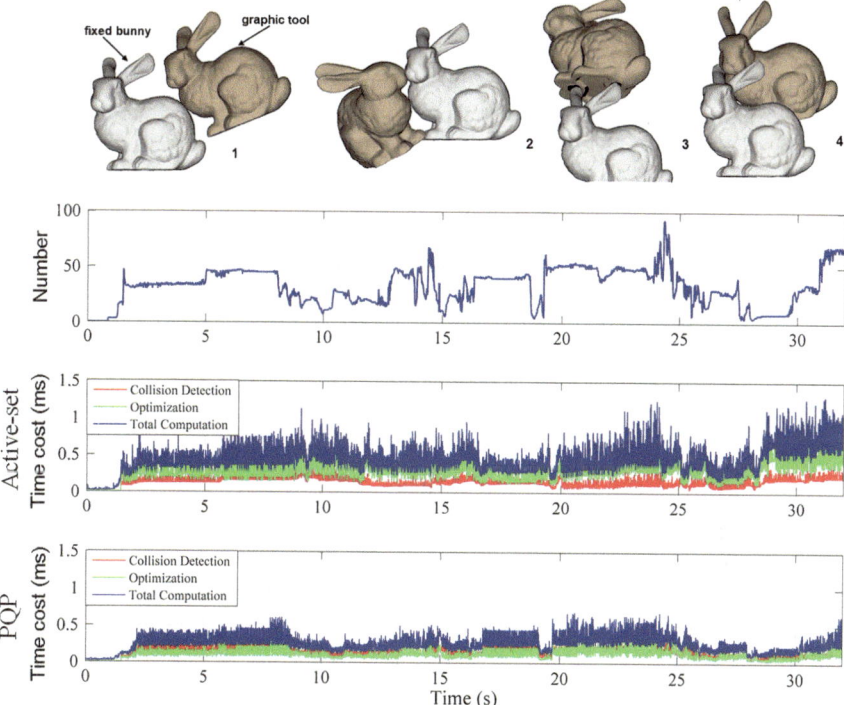

Fig. 3.24 Experimental results of haptic simulation between a pair of bunnies. Top: four steps in contact (between the golden bunny and the silver bunny). **a** The number of intersected pairs of spheres in each time step. **b** Time cost with the active-set-based method. **c** Time cost with the PQP-based method. Copyright © IEEE. All rights reserved. Reprinted, with permission, from Yu et al. (2013)

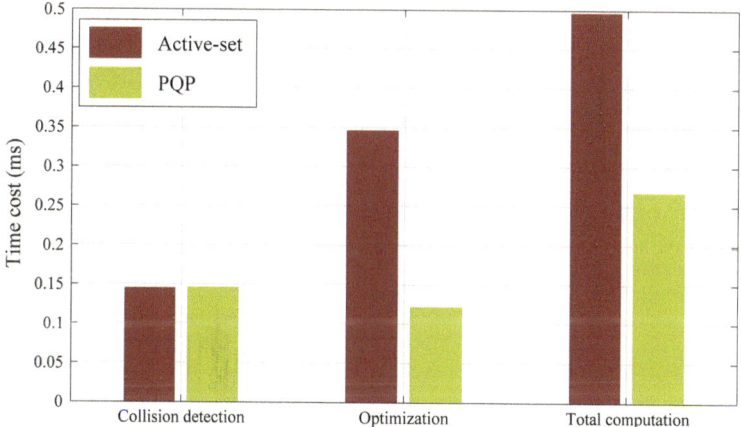

Fig. 3.25 The average time costs of collision detection, optimization, and the total computation with the two methods. Copyright © IEEE. All rights reserved. Reprinted, with permission, from Yu et al. (2013)

3.7 Summary

In this chapter, we have introduced a constraint-based 6-DoF haptic rendering algorithm for simulating haptic interaction involving sharp geometric features between two rigid bodies. A hybrid sphere-tree model has been described to represent an object with sharp edges for 6-DoF haptic interaction, and a threshold for a sphere's size has been introduced as a criterion to represent a sharp edge. Such a representation allows fast collision detection and small memory cost for simulation of sharp feature interaction. Based on the geometric representation, the configuration-based optimization approach is applied to solve the configuration of the graphic tool. The experiment results demonstrate that the method can reproduce the sharp feelings on sharp features with fast and stable haptic feedback.

A continuous collision detection method based on sphere trees is introduced to avoid the pop-through phenomenon during tool manipulation against small-sized objects (such as small calculi adhered to the surface of a tooth).

Furthermore, as an alternative to the active-set-based method for optimization, the PQP-based method is shown to be more efficient in handling large area contacts between two rigid bodies, which form a new 6-DoF haptic rendering solution as an alternative choice for *active-set*-based QP solver to accelerate optimization. It shows to be capable to meet both requirements of accuracy and fast in multi-region contact constraints.

For future research, more psychophysics experiments to verify a user's subjective perception should be conducted on more sophisticated sharp features, such as corners, darts, and cusps.

References

Boulch A, Marlet R (2012) Fast and robust normal estimation for point clouds with sharp features. In: Proceedings of ACM SIGGRAPH/Eurographics symposium. Geometry Processing (SGP'12), vol. 31(5), pp 1765–1774 Aug 2012

Brand M, Shilpiekandula V, Yao C (2011) A parallel quadratic programming algorithm for model predictive control. In: Proceedings of IFAC world congress, pp 1013–1039

Barbič J, James DL (2008) Six-DoF haptic rendering of contact between geometrically complex reduced deformable models. IEEE Trans Haptics 1(1):39–52

Chen YH, Lian LL, He XJ (2008) Haptic rendering of three-dimensional heterogeneous fine surface features. Comput-Aided Des Appl 5(1–4):1–16

Colgate JE, Stanley MC, Brown JM (1995) Issues in the haptic display of tool use. In: proceedings IEEE/RSJ international conference intelligence robots syststem, pp 140–145

Duriez C, Dubois F, Kheddar A, Andriot C (2006) Realistic haptic rendering of interacting deformable objects in virtual environments. IEEE Trans Vis Comput Graphics, 36–47

Gescheider G (1997) Psychophysics: the fundamentals, 3rd edn. Lawrence Erlbaum Associates Publisher, New Jersey

Hoppe H, DeRose T, Duchamp T (1994) Piecewise smooth surface construction. In: Proceedings of ACM SIGGRAPH, pp 295–302 (July 1994)

Lin MC, Otaduy M (2008) Haptic rendering: foundations, algorithms, and applications. A K Peters, Ltd

McNeely W, Puterbaugh K, Troy J, (1999) Six degree-of-freedom haptic rendering using voxel sampling. In: Proceedings of ACM SIGGRAPH

Milman R, Davison EJ (2008) A fast mpc algorithm using nonfeasible active set methods. J Optim Theor Appl 139:591–616

Ortega M, Redon S, Coquillart S (2007) A six degree-of-freedom god-object method for haptic display of rigid bodies with surface properties. IEEE Trans Vis Comput Graphics, 13 (3):458–469

Otaduy MA, Lin MC (2006) A modular haptic rendering algorithm for stable and transparent 6-DoF manipulation. IEEE Trans Robot 22(4):751–762

Richter S, Jones C, Morari M (2009) Real-time input-constrained MPC using fast gradient methods. In: Proceedings IEEE conference decision and control, Shanghai, pp 7387–7393

Shon Y (2006) Development and evaluation of a haptic rendering system for virtual design environments. PhD Dissertation, Engineering-Mechanical Engineering Department, University of California at Berkeley

Wang D, Zhang Y, Hou J, Wang Y, Lü P, Chen Y, Zhao H (2012a) iDental: a haptic-based dental simulator and its preliminary evaluation. In: Proceeding IEEE Transaction on Haptics, Volume: 5(4), pp. 332-343, Oct–Dec, 2012

Wang D, Liu S, Zhang X, Zhang Y, Xiao J (2012b) Six-degree-of-freedom haptic simulation of organ deformation in dental operations. IEEE international conference on robotics and automation, (ICRA), St. Paul, Minnesota, pp 1050–1056, 14–18 May, 2012

Wang D, Liu S, Xiao J, Hou J, Zhang Y (2012c) Six degree-of-freedom haptic simulation of pathological changes in periodontal operations. In: 2012 IEEE/RSJ international conference on intelligent robots and systems (IROS2012), Vilamoura, Algarve, 7–12 Oct, 2012

Wang D, Shi Y, Liu S, Zhang Y, Xiao J (2014) Haptic simulation of organ deformation and hybrid contacts in dental operations. IEEE Trans Haptics 7(1):48–60

Yamashita J, Lindeman RW (2000) On determining the haptic smoothness of force-shaded surfaces Conference abstracts and applications of ACM (SIGGRAPH'00), pp 240

Yamashita J, Fukui Y, Morikawa O (2001) Spatial resolution in haptic rendering. In: Proceedings of conference abstracts and applications of ACM (SIGGRAPH'01), pp 264

Yu G, Wang D, Zhang Y, Zhang X (2012) Six degree-of-freedom haptic simulation of sharp geometric features using a hybrid sphere-tree model. In: Proceedings IEEE/RSJ international conference on intelligent robots and systems (IROS2012), Vilamoura, Algarve, 7–12 October, 2012

Chapter 4
6-DoF Haptic Simulation of Deformable Objects

In this chapter, the configuration-based optimization approach for 6-DoF haptic simulation is extended to handle deformable objects and hybrid contacts. In Sect. 4.1, we introduce the problem and related work for haptic rendering of deformable objects. In Sect. 4.2, we provide an overview of the approach. In Sect. 4.3, we introduce the extended sphere-tree model with springs for deformable objects and the corresponding collision detection scheme. In Sect. 4.4, we extend the configuration-based method to deformation simulation. In Sect. 4.5, we propose an efficient method to simulate hybrid contacts, which are characterized by a tool interacting with both rigid and deformable objects, such as those between a dental probe and both a rigid tooth and its surrounding gingiva. In Sect. 4.6, we explain how to update a sphere tree under deformation. In Sect. 4.7, we describe optimization for determining the configuration of the graphic tool in contact and the corresponding contact force/torque. In Sect. 4.8, we present the results of applying the method to simulating dental operations. In Sect. 4.9, we conclude the chapter and discuss future work.

4.1 Related Work

Effective 6-DoF haptic rendering for deformable objects is an important topic that has many potential applications such as in virtual prototyping, surgical simulator for virtual training, and games (Lin and Otuday 2008; Wang et al. 2014). For example, as shown in Fig. 4.1, in a periodontal simulation, bimanual haptic interaction is needed to simulate calculus removal. A dentist (in training) uses the left hand to manipulate a dental mirror to push the tongue and expose the target tooth and the right hand to manipulate a dental probe in the periodontal pocket between the target tooth and its surrounding gingiva. During the insertion process, large deformation of the tongue and local deformation of the gingiva occur. The operator should feel

Electronic supplementary material The online version of this article (doi:10.1007/978-3-662-44949-3_4) contains supplementary material, which is available to authorized users.

D. Wang et al., *Haptic Rendering for Simulation of Fine Manipulation*, DOI 10.1007/978-3-662-44949-3_4

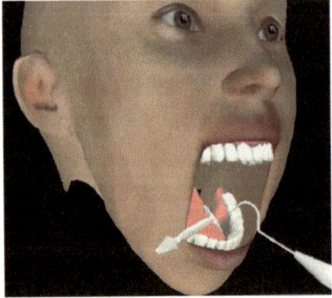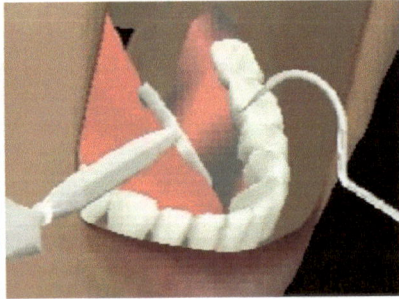

Fig. 4.1 Bimanual 6-DoF haptic interaction with deformation of the tongue and the gingiva. Copyright © IEEE. All rights reserved. Reprinted, with permission, from Wang et al. (2014)

force and torque when he/she pushes the tongue and moves the probe inside the periodontal pocket.

There has been considerable research on simulation of real-time interactions involving deformable objects (Nealen et al. 2006; Meier et al. 2005). Mass-spring-damper models were used in early systems due to their simplicity (Delingette et al. 1994; Zhong et al. 2005). A mass-spring model offers a simple and computationally efficient approach, but it is difficult to map actual material properties of the object into the model of spring and damper (Shi and Payandeh 2010). More accurate constitutive models of soft object deformation based on continuum mechanics have also been proposed (Bro-Nielsen 1998; Wu et al. 2001). Compared with mass-spring models, finite element methods can produce physically accurate behaviors. However, due to the large size of the system matrix and the complexity of shape functions, these methods have high computational costs and are difficult to use for real-time haptic interaction (Shi and Payandeh 2010).

Among earlier publications on haptic rendering of deformable objects, most of them treat the tool as a point and simulate point or single-region contact between the tool and a deformed object (Conti et al. 2003; Zhuang and Canny 2000).

In recent years, haptic rendering of multi-region contacts involving deformable objects has drawn much attention. Barbič and James introduced a time-critical method for 6-DoF haptic rendering of deformable objects (Barbič and James 2008). They used a penalty-based method for computing the contact force and virtual coupling for maintaining the stability of force rendering. Luo and Xiao introduced a beam skeleton to simulate deformation under multi-region contacts (Luo and Xiao 2007). Penalty-based methods cannot prevent perceptible penetrations when simulating exact contacts among objects. Contact force/torque rendering based on penalty methods may result in instability; therefore, virtual coupling is usually used to ensure stability at the cost of decreasing force/torque fidelity.

Different from the penalty-based approach, the constraint-based approach aims to constrain the pose of the graphic tool to be free of penetration, while the haptic tool can penetrate into objects (Wang et al. 2013). Duriez et al. introduced a constraint-based method for simulation of deformation due to contacts (Duriez et al. 2006;

Courtecuisse et al. 2010). While multi-region contacts were modeled accurately, the update rate was below 100 Hz even after using GPU-based parallel computation and a layered depth images (LDI) approach for accelerating contact detection (Courtecuisse et al. 2010). Recently, Garre et al. (2011) modeled the deformation of a human hand upon interacting with rigid objects by using a skeleton–skin hybrid model. Existing constraint-based methods are computationally expensive, and the update rate of deformation is dependent on the number of DoFs, which is usually about 30–60 Hz. Therefore, Peterlik et al. (2011) utilized a multi-rate compliant mechanism to update the contact constraints between two deformation loops in order to compute forces in a 1 kHz haptic update rate.

Several approaches were reported on dental simulation involving haptics (Wang et al. 2009; Luciano et al. 2009; Tse et al. 2010; Forsslund et al. 2009). Only a few of them provide deformation simulation of oral organs. The Simodont system considered small, local deformation of the tongue.[1] Large deformation of the tongue (for exposing the target teeth) was not addressed. Furthermore, there is no model to simulate *hybrid contacts*, i.e., simultaneous contacts and force/torque feedback between dental probe and both a rigid tooth and a deformed gingiva as the operator inserts a dental probe into the narrow periodontal pocket.

4.2 Overview of the Approach

We now provide the framework of the approach and explain unique features compared to previous 6-DoF haptic rendering approaches for deformation simulation. In the simulation, we assume that the tool is rigid, but the objects (e.g., organs) can be elastic. The main ideas of the approach include:

(1) Use a sphere-tree model for the tool and a sphere-tree model with springs for each deformable object, and compute deformation by updating the sphere tree efficiently based on the contact force.
(2) Apply the configuration-based optimization method to the updated object model to compute the rigid tool configuration to avoid penetration between the graphic tool and the deformed objects.

The flowchart of the approach is shown in Fig. 4.2. In every haptic simulation cycle, first, the configuration of the haptic tool is obtained. Then, collision detection based on sphere trees is carried out to detect pairs of intersected spheres at each contact region.

Next, the spheres connected by springs, called a skeleton, are altered to model large deformation, and a geometry-based method is used to compute local deformation. Once deformation is computed, the hierarchy between parents and their children spheres in a sphere tree is updated accordingly.

---

[1] http://www.moog.com/markets/medical-dental-simulation/haptic-technology-in-simodont/

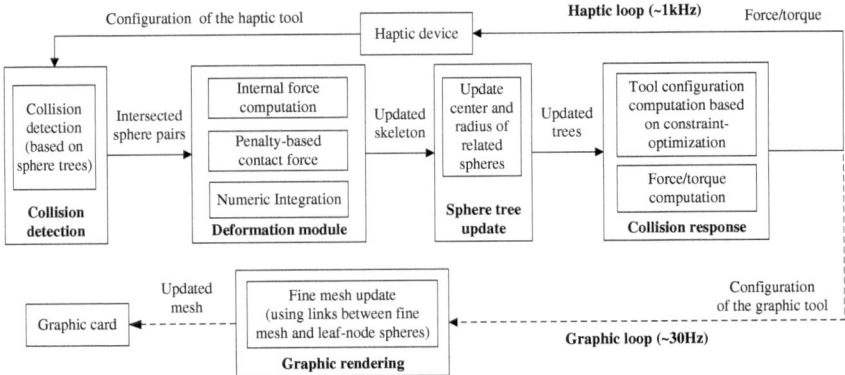

Fig. 4.2 Flowchart of the 6-DoF haptic rendering method involving deformable objects.
Copyright © IEEE. All rights reserved. Reprinted, with permission, from Wang et al. (2014)

For accurate collision response, the configuration-based optimization method
(Chap. 2) is used to compute a configuration for the graphic tool contacting the
deformed objects with no penetration and to compute the six-dimensional force/
torque response to the haptic tool.

For graphic display, we use a fine triangle mesh for a deformed object. Each
sphere of the object's sphere-tree model is associated with multiple vertices of the
mesh model. Based on the updated sphere tree, the positions of all associated
vertices on the mesh are also updated in the relatively slow graphic loop.

The proposed approach provides a new, unified solution for 6-DoF haptic
simulation involving deformable objects in *multi-region and hybrid contacts*, which
are characterized by the following features:

(1) *Efficient, unified object model*:

The approach uses a sphere tree to represent a general rigid object and a sphere
tree with springs to represent a general deformable object and conducts all related
computation accordingly (including collision detection, deformation simulation,
collision response for the tool, and haptic force/torque simulation). It takes full
advantage of the simplicity of spheres to be efficient in computation.

In contrast, multi-representation models were used in previous work of both
penalty-based and constraint-based methods: Barbič and James used a point-shell
model for collision detection and a point-based modification of the bounded
deformation tree (BD-Tree) for updating the point-shell model (Barbič and James
2008); Mendoza and O'Sullivan used a sphere tree for collision detection and a
reduced model (e.g., a coarser mesh) for deformation computation. Such multiple
representations for each deformable object incur the extra cost of synchronizing
computations for different types of models in each simulation cycle (Mendoza and
O'Sullivan 2005).

(2) *Single thread simulation*:

With the unified object representation, the approach enables a single thread at the haptic rate (i.e., 1 kHz) for all relevant computations involved. In contrast, existing constraint-based methods used multiple simulation threads: Deformation and dynamics computations were carried out in a low-rate simulation loop, and force computation was carried out in a high-rate haptic rendering loop (Duriez et al. 2006; Courtecuisse et al. 2010; Peterlik et al. 2011). Additional intermediate models such as the multi-rate compliant mechanisms have to be computed twice and shared by the two loops. Such asynchronous update of the constraint information can lead to loss of the subtle force feeling in force feedback, especially when the tool slides along certain fine features of an object.

(3) *Decoupled computation of deformation and tool optimization*:

Being constraint based, the approach avoids interpenetration between a rigid tool and either a deformable object or a rigid object. In each simulation step, it first computes the deformed configuration of an object (e.g., a tongue or gingiva), and then, based on this deformed configuration, compute the configuration of the graphic tool. With this approach, the optimization of the graphic tool's configuration can rely on fast existing methods (Wang et al. 2013). The method can maintain hybrid contacts between a complex shaped tool and both a rigid and a deformable object without penetration. A few researchers addressed such hybrid contacts (Garre and Otaduy 2010), but more rigorous investigation is required to enable high-fidelity simulation of fine manipulations.

4.3 Modeling Using Sphere-Trees with Springs

We represent objects based on the sphere-tree model and extend the representation to a sphere-tree model with springs (for deformable objects), in order to take full advantage of the simplicity of spheres to achieve high efficiency and the optimization-based collision response method (see Chap. 2). After all, in applications such as dental simulation, it is not a high priority to accurately model sophisticated nonlinear viscoelasticity of the tongue or gingiva under the external force, but it is more important to provide proper deformation and contact response without time delay.

We extend a sphere tree by adding springs and dampers at a high level (such as level 2) to model a deformable object. As shown in Fig. 4.3, the spring-connected high-level spheres form a skeleton for the sphere tree. The elastic connection between two adjacent spheres in the skeleton is similar to that used in Conti et al. 2003. This skeleton can be deformed by compressing or extending the springs to simulate large deformation.

Furthermore, a time-critical method using multi-levels of skeletons can be introduced to achieve trade-offs between accuracy and time cost. During each haptic simulation cycle, we first use level 2 (sixty-four spheres) as the skeleton of

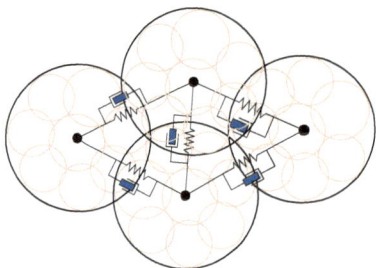

Fig. 4.3 A sphere tree with springs and dampers in the skeleton. Copyright © IEEE. All rights reserved. Reprinted, with permission, from Wang et al. (2014)

the deformed object. If sufficient time is left in the current simulation cycle, we use a finer level of the sphere tree to achieve higher spatial resolution for the deformation simulation.

The skeleton model is flexible in that it can be used to model several properties of organs. First, the spring stiffness along the longitude and latitude directions of an object (such as a tongue) can be different to simulate non-isotropic deformation. Second, a nonlinear force deformation profile can be defined with changing stiffness. Third, non-homogenous properties can be simulated, for example, by setting large stiffness at the base of the tongue and small stiffness at the tip.

Another advantage of the skeleton model is that it can be optimized to achieve fairly accurate deformation results compared to its finite element model (FEM) counterparts. By constructing a similarity metrics at several sampling points (e.g., centers of all the spheres in the skeleton), we can compute the geometric error based on the deformation results from the two methods as a feedback signal to adjust the value of the skeleton parameters and reduce the error to an acceptable level.

4.4 Contact Constraint Prediction Method

We now consider detection of intersections between the sphere tree of a rigid haptic tool and the sphere tree of a deformable object when it is undeformed.

In order to speed up the computation, we introduce a contact constraint prediction (CCP) method to eliminate the number of intersecting spheres that are redundant for forming the non-penetration contact constraints. The CCP method takes advantage of the graphic tool's contact configuration in the previous time instant $t - 1$. We first decide the maximum distance δ that the tool can move in the interval between $t - 1$ and t based on the speed of the haptic tool (which is assumed to be 0.5 mm/ms according to a person's average moving speed). Next, we grow each sphere of the graphic tool at $t - 1$ by δ and find the pairs of intersecting spheres between the (enlarged) graphic tool and the object. We use those intersecting sphere pairs to form the non-penetration contact constraints and compute

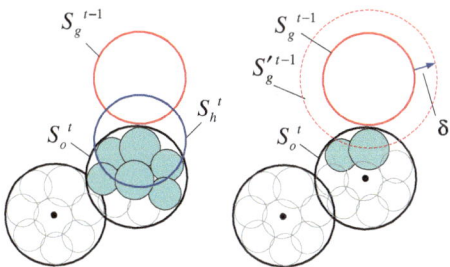

Fig. 4.4 CCP reduces the number of constraints (the *left figure* illustrates intersecting spheres by using the *blue* haptic tool; the *right figure* illustrates the intersecting spheres by using the enlarged *red* graphic tool). Copyright © IEEE. All rights reserved. Reprinted, with permission, from Wang et al. (2014)

the penetration depth between the two intersecting spheres. The essence is that instead of directly checking intersection between the haptic tool and the objects at t, the CCP method checks intersections between the grown graphic tool at $t-1$ and the objects, which are far fewer.

As shown in Fig. 4.4, the CCP method greatly reduces the number of constraints without missing necessary constraints. S_g^{t-1} denotes a sphere from the graphic tool at the previous time step, $S_g'^{t-1}$ denotes its enlarged version, S_h^t denotes a sphere from the haptic tool at the current time step, and S_O^t denotes a sphere from the deformed object before its deformation.

The above method for constraint reduction can greatly reduce the contacting sphere pairs considered in the optimization process and thus reduce the time cost of the optimization. Given a value of δ and geometrical information of the objects, an upper bound n on the number of contacting sphere pairs per contact region could be determined. In the implementation of the method, we can tune the value of δ to restrict the value of n (e.g., n is less than 10). Suppose there are less than 50 simultaneous contact regions in a multi-region contact scenario, the maximum number of constraint inequalities from contacting sphere pairs is thus 500, which can be easily solved within 1 ms.

4.5 Skeleton-Based Deformation Computation

We simulate large, global deformation of an object through deforming the springs in the skeleton of the object's sphere tree. As introduced in Sect. 4.3, the skeleton model is flexible to model several properties of organs and provides a good compromise between accuracy and time efficiency. The algorithm for computing large deformation consists of the following steps:

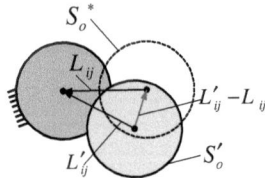

Fig. 4.5 Illustration of the internal force. Copyright © IEEE. All rights reserved. Reprinted, with permission, from Wang et al. (2014)

1. Compute internal force for each sphere in the skeleton of the object;
2. Compute external force for each contact sphere in the skeleton of the object;
3. Formulate three-dimensional Newtonian equations for each sphere of the object, and integrate the equations to solve for the updated centers of all the spheres in the skeleton.

4.5.1 Computation of the Internal Force

The internal force on each ith sphere of the deformable object can be computed as

$$\mathbf{F}_i^{\text{int}} = \sum_{j=1}^{n_i} k_s \left(\mathbf{L}'_{ij} - \mathbf{L}_{ij} \right) - B_{ci}\mathbf{v}_i \tag{4.1}$$

where k_s is the stiffness of each spring. n_i is the number of springs attached to the ith sphere, $\mathbf{L}_{ij} = \mathbf{C}_i - \mathbf{C}_j$ is the distance vector between the center $\mathbf{C}_i$ of the ith sphere and the center $\mathbf{C}_j$ of an adjacent jth sphere when the object is not deformed, and $\mathbf{L}'_{ij}$ is the corresponding distance vector between the two adjacent spheres when the object is deformed. As shown in Fig. 4.5, S_o^* indicates the ith sphere from the undeformed object, and S'_O is the corresponding sphere when the object is deformed. $\mathbf{v}_i$ is the velocity of the sphere S'_O (which will be computed by numeric integration), and B_{ci} is the damping coefficient, which will be explained in Sect. 4.5.4.

4.5.2 Computation of the External Force

In order to compute resisting force from the deformable objects, a straightforward and most common method is to use the distance between the haptic tool and the deformable objects (Lin and Otaduy 2008). However, as shown in Fig. 4.4, the problem of this method is that it will produce too many penetrated pairs of spheres when the haptic tool penetrates deeply into the surface of the deformable object,

which is a common case when the operator exerts a large force. Obviously, the
number of penetrated pairs of spheres will directly influence the computation cost
of the external force and subsequent object deformation. Even worse, when the
haptic tool is fully immersed into the deformed object, some contact spheres will be
missed by collision checking at discrete configurations between the haptic tool and
the deformed object. In order to solve this problem, we introduce a new method to
detect all the contact spheres at the surface of the deformable object and, at the same
time, decrease the number of spheres needed for external force computation.

If the haptic tool moves toward the object from time $t - 1$ to t, then we use the
CCP method (Sect. 4.4) to find the pairs of contact spheres between the graphic tool
and the deformable object. No matter if the object is undeformed or already
deformed at the end of $t - 1$, we determine the external force to a contact sphere of
the object based on the distance between the graphic tool at time step $t - 1$ and the
haptic tool at time step t. As shown in Fig. 4.6, S_g^{t-1} denotes a sphere from the
graphic tool at time $t - 1$, and S_h^t denotes the corresponding sphere from the haptic
tool at time t. Figure 4.6a, b denotes the cases when the object is undeformed and
deformed at the end of time step $t - 1$, respectively. In either case, the external
force at the beginning of time step t can be computed as

$$\mathbf{F}_i^{\text{ext}} = K_p \delta_i^p \tag{4.2}$$

Fig. 4.6 Computation of the
penetration vector at time step
t. Copyright © IEEE. All
rights reserved. Reprinted,
with permission, from Wang
et al. (2014). **a** Object
undeformed at the end of time
step $t - 1$. **b** Object deformed
at the end of time step $t - 1$

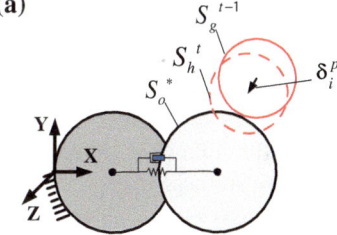

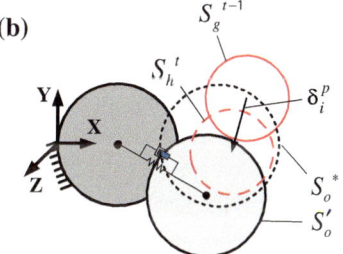

where K_p is the penalty stiffness and the vector $\boldsymbol{\delta}_i^p$ is defined and computed as follows:

1. Find the contact sphere S_g^{t-1} of the graphic tool and the corresponding sphere S_h^t of the haptic tool.
2. Denote the centers of the S_g^{t-1} and S_h^t as $\mathbf{C}_{gi}$ and $\mathbf{C}_{hi}$, respectively.
3. Compute vector $\boldsymbol{\delta}_i^p$ as

$$\boldsymbol{\delta}_i^p = \mathbf{C}_{hi} - \mathbf{C}_{gi} \tag{4.3}$$

Note that if the haptic tool is out of contact at t, Eq. (4.2) cannot be used, and the external force is set to zero.

4.5.3 Computation of Deformation

Based on Eqs. (4.1) and (4.2), the following Newtonian equations are formulated for each ith sphere of the object to solve for the position $\mathbf{X}_{ci}$ of the deformed sphere

$$m\ddot{\mathbf{X}}_{ci} = \mathbf{F}_i^{\text{ext}} + \mathbf{F}_i^{\text{int}} \tag{4.4}$$

where m is the mass of the ith sphere in the skeleton (and we assume that the mass of a sphere is concentrated at its center).

4.5.4 Stability Analysis

The values of the following parameters will affect the stability of the deformation process: penalty stiffness matrix has eigenvalues K_p, spring stiffness K_{Si}, damping coefficient B_{ci}, mass m of each sphere, number of springs attached to each sphere, integration time step Δt, etc.

In order to validate whether the decoupled computation between deformation and tool optimization can maintain stability of the haptic simulation, we used a 1-DoF system as a benchmark to test the convergence of the deformation computation, as well as the capability of producing valid optimization results and continuous change of the feedback force.

As shown in Fig. 4.7, a simplified deformable object has to consist of at least two spheres. The shaded sphere is fixed, and the black sphere can move along the x-axis to deform the spring. The red dashed sphere represents the haptic tool, and the red sphere represents the graphic tool. The trajectory of the haptic device is constrained along the x-axis.

Fig. 4.7 Deformation
analysis by using a 1D
example. Copyright © IEEE.
All rights reserved. Reprinted,
with permission, from Wang
et al. (2014). **a** No
deformation at the beginning
of $t-1$. **b** Deformation
occurred at the beginning of t

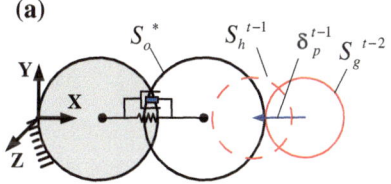

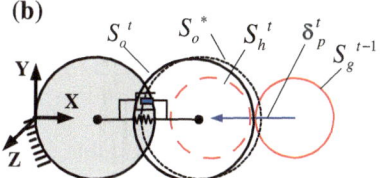

   To compute the updated center of each sphere in the object skeleton, we use the
explicit Euler integration which is easy to implement. To ensure that the integration
converges to the equilibrium solution, we need to select a suitable value for the
integration time step Δt.

   Based on the Courant condition (Fierz et al. 2011; Müller et al. 2005), we now
derive the stable range of the time step for damped mass-spring systems. We
consider a linear damped mass-spring system to explain the stability of explicit
integration. Based on Eq. (4.4), the next-step velocity $v(t+\Delta t)$ and displacement
$x(t+\Delta t)$ of the displaced sphere can be derived as

$$v(t+\Delta t) = v(t) + \Delta t \frac{k_p[x_h - (x(t)+r_h+r_o)] - k_S(x(t)-l_0) - b_c v(t)}{m}$$
$$x(t+\Delta t) = x(t) + \Delta t \cdot v(t) \tag{4.5}$$

where Δt is the time step. k_p is the penalty stiffness to compute the external force. x_h
is the position of the center of the haptic tool. r_h is the radius of the haptic tool. r_o is
the radius of the displaced object sphere. k_S and b_c are the spring and damping
coefficient, respectively. l_0 is the rest length of the spring, and m is the mass of the
displaced object sphere.

   The matrix form of Eq. (4.5) is

$$\begin{pmatrix} v(t+\Delta t) \\ x(t+\Delta t) \end{pmatrix} = \begin{pmatrix} 1 - \frac{b_c \Delta t}{m} & -\frac{(k_S+k_p)\Delta t}{m} \\ \Delta t & 1 \end{pmatrix} \begin{pmatrix} v(t) \\ x(t) \end{pmatrix} + \begin{pmatrix} \frac{\Delta t l_0}{m} \\ 0 \end{pmatrix} \tag{4.6}$$

The system matrix has eigenvalues

$$\lambda = 1 - \frac{\Delta t}{2m}\left(b_c \pm \sqrt{b_c^2 - 4(k_S + k_p)m}\right) \tag{4.7}$$

To ensure the stability of the system, the spectral radius of the system matrix, $\rho(A) := \max(\|\lambda_0\|, \|\lambda_1\|)$, should not be greater than one (Fierz et al. 2011; Müller et al. 2005), that is,

$$\left\| 1 - \frac{\Delta t}{2m}\left(b_c \pm \sqrt{b_c^2 - 4(k_S + k_p)m}\right) \right\| \le 1 \tag{4.8}$$

Hence, we obtain the stable range of the time step Δt:

$$\Delta t \le \Delta t^* = \begin{cases} \frac{b_c}{(k_S + k_p)} & \rho < 0 \\ 2\sqrt{\frac{m}{(k_S + k_p)}} & \rho = 0 \\ \frac{4m}{b_c + \sqrt{\lambda}} & \rho > 0 \end{cases} \tag{4.9}$$

where

$$\rho = b_c^2 - 4(k_S + k_p)m \tag{4.10}$$

Equation (4.9) provides the relationship between the time step and the stiffness of the deformable environment to maintain the stability of the deformation simulation. With Eq. (4.9), we can tune the parameters to simulate stiff or soft deformable effects while maintaining convergence of the integration and the stability of the simulation. In the simulation, the effective time step Δt is always set as the time period for one haptic simulation cycle, i.e., 1 ms. This time step should satisfy Eq. (4.9).

4.6 Local Deformation and Hybrid Contacts

To simulate fine manipulation, such as inserting a dental probe into a narrow periodontal pocket, there are two challenges need to be addressed:

1. The trade-off between efficiency and high accuracy in simulating local deformation (such as the deformation of a gingiva);
2. Fidelity and stability of fine manipulation within a narrow space where the tool forms simultaneous contacts with both rigid and deformed objects (i.e., a tooth and a periodontal pocket), which we call *hybrid contacts*.

The fidelity of local deformation simulation is crucial to the outcome of fine manipulations. In this section, we use gingiva as an example to study the simulation algorithm for local deformation and hybrid contacts. The proposed algorithm could be extended to other scenarios by tuning the physical parameters of the deformed objects.

4.6.1 Local Deformation of Gingiva

As the deformation of a gingiva is always small and local, a neighborhood of each contact point will be identified so that the spheres outside of this neighborhood will not be affected by deformation. As the contact point changes, the corresponding neighborhoods change. Local deformation is simulated by changing the center position of each sphere within the neighborhood that intersects with the sphere tree of the rigid tool, which we call fringe spheres.

To simulate the small clearance formed by a periodontal pocket, we use a sphere tree of deeper levels to model the gingiva and use a finer level (e.g., level 3) for deformation computation in order to limit the geometric error between the sphere-tree representation and the corresponding triangle mesh representation for graphic display.

During the off-line construction of the sphere tree, we carry out a manual segmentation of the gingiva of the lower jaw into several sub-gingiva (shown in Fig. 4.8a), i.e., for each tooth, we define its surrounding gingiva. Each sub-gingiva is modeled by a deeper sphere tree (e.g., with three levels).

The Hook's law is used to compute a displacement (such as $\boldsymbol{\delta}_{s0}$ in Fig. 4.9) of the center of each fringe sphere (such as S_o in Fig. 4.9) with the following equation:

$$\begin{cases} \boldsymbol{\delta}_{s0} = \frac{\mathbf{F}_{ext} \cdot \mathbf{n}_{inward}}{k_d} \cdot \mathbf{n}_{inward} & \mathbf{F}_{ext} \cdot \mathbf{n}_{inward} \geq 0 \\ \boldsymbol{\delta}_{s0} = 0 & \mathbf{F}_{ext} \cdot \mathbf{n}_{inward} < 0 \end{cases} \tag{4.11}$$

(a)   **(b)**

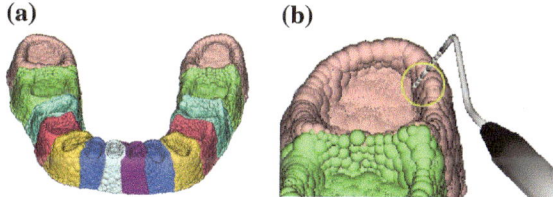

Fig. 4.8 Illustration of local deformation method on a gingiva model. Copyright © IEEE. All rights reserved. Reprinted, with permission, from Wang et al. (2014). **a** Segmentation of the gingiva into several sub-gingivas and **b** determination of an active neighbor region (the *yellow circle*)

Fig. 4.9 Local deformation

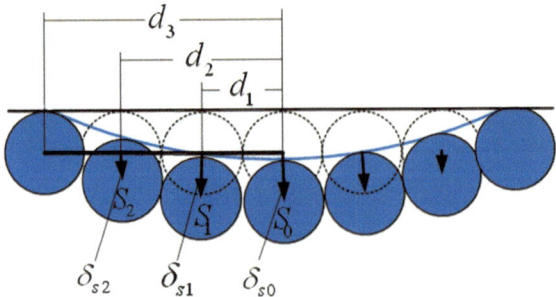

where k_d denotes the spring stiffness that can be tuned to reflect different organ property. $\mathbf{F}_{ext}$ is computed by

$$\mathbf{F}_{ext} = K_p \cdot \boldsymbol{\delta}_i^p \tag{4.12}$$

where $\boldsymbol{\delta}_i^p$ refers to the penetration vector computed by Eq. (4.3). $\mathbf{n}_{inward}$ denotes the allowed direction of local deformation, which is defined as follows:

$$\mathbf{n}_{inward} = \frac{\vec{c_2 c_1}}{\|\vec{c_2 c_1}\|} \tag{4.13}$$

where c_1 denotes the center of the displaced sphere on the gingiva and c_2 denotes the center of a closest sphere on the tooth to the displaced sphere (as shown in Fig. 4.10).

After the fringe spheres are displaced (i.e., their center positions are changed), we update the centers of neighboring spheres of each fringe sphere according to a distance-based interpolation function, such as an exponential attenuation function, as the following:

$$\boldsymbol{\delta}_{si} = \alpha^{d_i} \boldsymbol{\delta}_{s0} \tag{4.14}$$

where the parameter α has the range of (0, 1), which is set as a constant value for all spheres. The greater the value of α, the closer is the displacement $\boldsymbol{\delta}_{si}$ of sphere S_i to

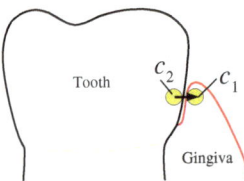

Fig. 4.10 Allowed deformation direction. from Wang et al. (2014)

$\delta_{s0} \cdot d_i$ is a power of α, which is the distance between sphere S_i (e.g. S_1 or S_2) and the fringe sphere S_o as shown in Fig. 4.9.

4.6.2 Simulation of Hybrid Contacts

Force fidelity is important for correct measurement of a pocket's depth. There are two aspects of the interaction force during the probe's motion within the pocket.

First, a resistance force threshold at the bottom of the periodontal pocket needs to be simulated. When the probe approaches the bottom of the pocket, the gingiva will be damaged if the exerted force along the vertical direction is greater than a predefined threshold (e.g., 20 g). In the method, we define a target depth for each pocket and only allow the spheres above the bottom of the pocket to be displaced.

Second, responsive changes of force/torque caused by switches of contact states need to be simulated in periodontal operations, where the probe operates in the narrow space of the pocket between a tooth and its gingiva and can cause frequent

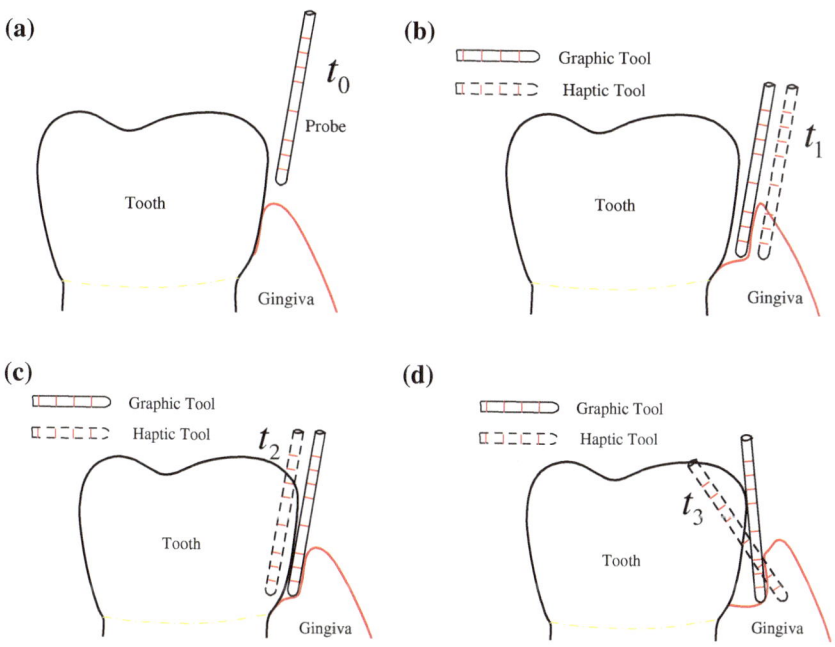

Fig. 4.11 Responsive force/torque change due to switches of contact states. Copyright © IEEE. All rights reserved. Reprinted, with permission, from Wang et al. (2014). **a** Before deformation. **b** Lateral force from only gingiva. **c** Force from both tooth and gingiva **d** Force and torque

changes of contact states (shown in Fig. 4.11). Frequent contact switches can cause penalty-based rendering unstable.

With the configuration-based optimization method, a non-penetrating configuration of the graphic tool is computed for each contact state, which can maintain stability under frequent contact switches. We extend this approach to narrow spaces resulted from simultaneous multi-region, hybrid contacts between the tool and both rigid and deformable objects (e.g., the rigid tooth and its deformable gingiva). We discuss the new challenges and present the approach below.

As shown in Fig. 4.11b, c, when the haptic tool moves from time t_1 to t_2, a hybrid contact between the tool and both the rigid tooth and the deformable gingiva is formed. Vibration can occur with the deformable model in Eq. (4.11). This is because the local deformation displacement δ_{s0} is zero, which releases the gingiva back to its undeformed position. Without deformation, there is no space for the tool, and thus, the graphic tool computed by the subsequent optimization module will jump from inside to outside of the pocket.

In order to prevent the vibration that can occur with hybrid contacts, we propose a modified deformation model to ensure that when hybrid contacts occur, the gingiva cannot fully go back to its undeformed position even if the haptic tool is not penetrating the gingiva, because the graphic tool needs to stay within the pocket.

It is not straightforward to ensure the above stability requirements. In the modified deformation model, the history of contacts and the position of the haptic tool at the current time step are both considered and possible interaction types are handled.

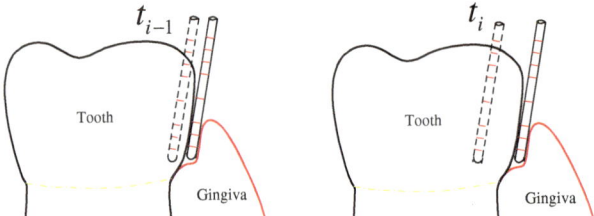

Initial condition: haptic tool penetrates tooth at t_{i-1} Type 1: haptic tool penetrates tooth at t_i

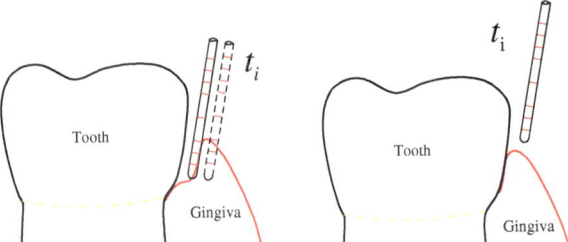

Type 2: haptic tool penetrates gingiva   Type 3: no penetration

Fig. 4.12 Three interaction types at current time step t_i. Copyright © IEEE. All rights reserved. Reprinted, with permission, from Wang et al. (2014)

If hybrid contacts did not occur at the previous time step, the deformation of gingiva at the current time step is computed by Eq. (4.11).

If hybrid contacts occurred at the previous time step t_{i-1}, we define three interaction types as illustrated in Fig. 4.12. Local deformation of the gingiva at the current time step t_i is computed as follows:

$$
\begin{cases}
\delta_{s0} = 2R \cdot \mathbf{n}_{inward} & (\mathbf{F}_{ext} \cdot \mathbf{n}_{inward} \leq 0) \\
\delta_{s0} = \frac{\mathbf{F}_{ext} \cdot \mathbf{n}_{inward}}{k_d} \cdot \mathbf{n}_{inward} & (\mathbf{F}_{ext} \cdot \mathbf{n}_{inward} > 0)
\end{cases}
\tag{4.15}
$$

where R denotes the radius of the tip of the probe.

The first equation in Eq. (4.15) simulates type 1, where the gingiva is released back to an intermediate position instead of going back to the undeformed position, while the graphic tool is simultaneously contacting both the gingiva and the tooth. The second equation in Eq. (4.15) simulates type 2, where the deformation of the gingiva increases to a larger value.

For type 3, the deformation of the gingiva is computed as

$$
\delta_{s0} = 0
\tag{4.16}
$$

that is, the gingiva is released back to its undeformed position.

It should be noted that the method in this section is suitable for hybrid contacts in other applications, such as closely located bones and deformable organs in laparoscopic surgeries (Peterlik et al. 2011; Saupin et al. 2008).

4.7   Update of Sphere-Tree Model After Deformation

After computing deformation in terms of new centers of skeleton spheres for large deformation and fringe spheres for small deformation, we need to update the object's sphere tree to maintain proper parent/child relationship in the tree.

There are two kinds of update: from child nodes to parent nodes and from parent nodes to child nodes. Given the new positions of several children spheres (due to deformation computation), the radius of the parent sphere is updated so that the parent sphere can still contain those children spheres, as shown in Fig. 4.13. This process will continue until the root level is reached.

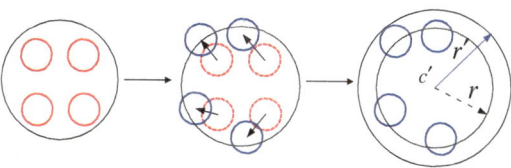

Fig. 4.13 Update of a parent sphere according to the new positions of its children. Copyright ©

The update rule is as follows:

$$\begin{cases} c' = c \\ r' = \max(|\overline{cc_i}| + r_i)_{i=1,\dots,8} \end{cases} \tag{4.17}$$

where c' and r' denote the new center and radius of the parent sphere, respectively. c denotes the old center of the parent sphere, and c_i and r_i denote the center and radius of the children spheres.

The update from parent nodes to child nodes is relatively simple. If the center position of a parent sphere is shifted by the deformation, its children spheres are shifted with the same displacement as the parent sphere.

4.8 Performance Analysis and Experimental Results

In the experiments, a Phantom Premium 3.0 6-DoF is utilized as the haptic device to provide six-dimensional forces/torques, and the computer used has the following specification: Intel(R) Core(TM) 2 2.20 GHz, 2 GB memory, and X1550 Series Radeon graphic card. Note that no GPU is used.

4.8.1 Performance of Global Deformation

In a typical periodontal operation, the target tooth is usually occluded by the tongue. A dentist uses a dental mirror to push and deform a tongue so that the target tooth is exposed for observation. During the operation, global deformation of the tongue will occur.

As shown in Table 4.1, several combinations of parameters are provided to illustrate the simulation performance of the 1-DoF deformation simulation. A video is provided to show the real-time simulation effects for each simulation case (the video is available online from the book Website http://extras.springer.com/). It illustrates that the method could simulate either large or small deformation effects

Table 4.1 Different simulation cases ($m = 20\,\text{g}$, $\zeta = 0.5$, $k_t = 200\,\text{N/m}$, $k_p = 5\,\text{N/m}$)

Case	Δt^*	k_S (N/m)	Deformation	Performance
C1	0.02	0.5	Large deformation	Stable
C2	0.03	2.5	Medium deformation	Stable
C3	0.011	150	Small deformation	Stable
C4	0.01	185	Overflow	Unstable
C5	0.0006	$5 \times 1e-4$	Delayed deformation	Unstable

while maintaining stability. It also illustrates that the interaction was stable under various contact velocities. The effects under different damping ratios are also shown.

The above testing shows that the effective way to realize large deformation and stable force feedback is to use a large coupling stiffness k_t (cf. defined in Eq. 2.29) in the stable impedance range of the haptic device (e.g., smaller than 0.9 N/mm for the Phantom Premium 3.0 6-DoF) and a small ratio for k_s/k_p. The smaller the ratio, the greater is the deformation. Furthermore, the upper bound for k_S should be set properly to prevent the actual time step Δt greater than the allowable time step limit Δt^*, which will violate the stability condition. In another aspect, for a given k_S, if we wish to increase the time step limit Δt^*, we may use a greater value of m to simulate a larger mass.

Besides validating the stability of deformation, the experiment results show that the method could produce valid optimization results and maintain continuous

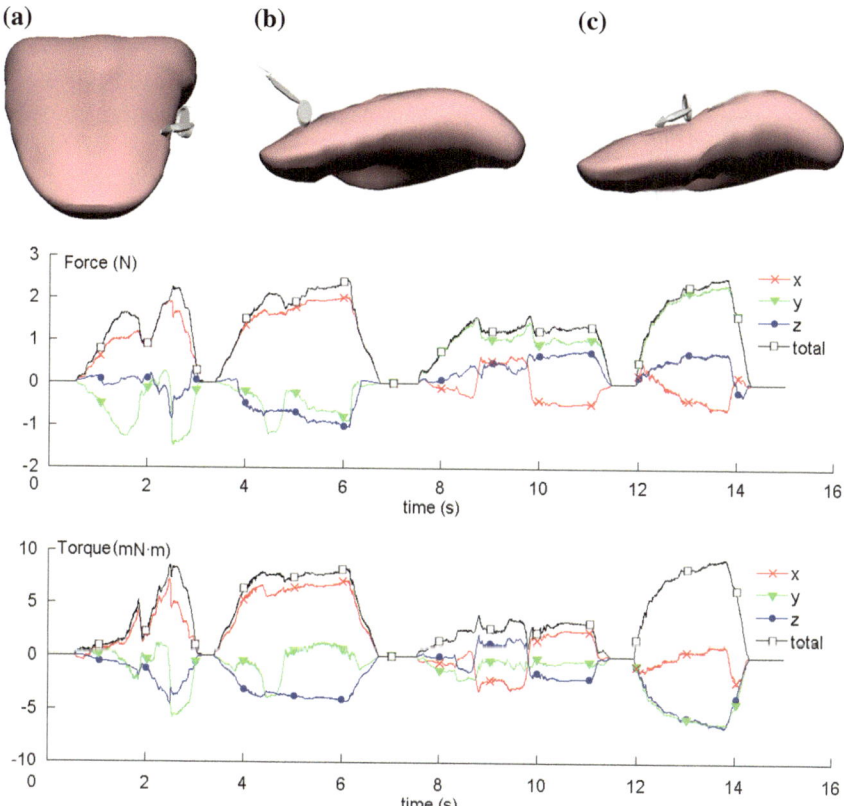

Fig. 4.14 Responsive deformation, consisting of three states: **a** local deformation only, **b** global deformation, and **c** deformation from multi-region contacts. Copyright © IEEE. All rights reserved. Reprinted, with permission, from Wang et al. (2014)

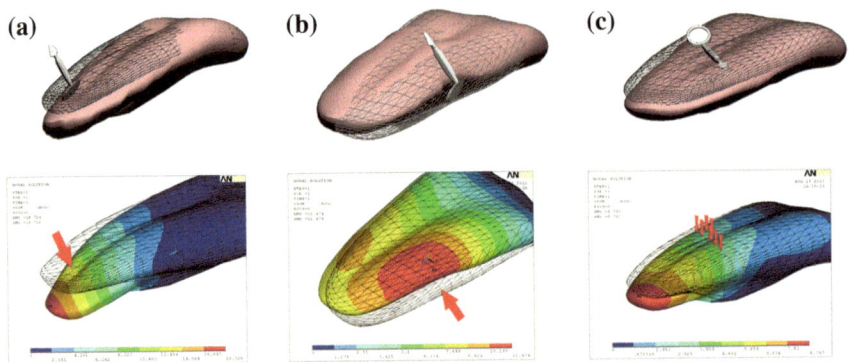

Fig. 4.15 Deformation results by the method (*upper figures*) and by ANSYS (*lower figures*): **a** vertical deformation, **b** lateral deformation, and **c** deformation under multiple contact regions. Copyright © IEEE. All rights reserved. Reprinted, with permission, from Wang et al. (2014)

change of the feedback force, which can be observed from the video. The user can feel stable force feedback under either large or small collision velocities.

Figure 4.14 presents some results of haptic simulation of a tool interacting with a tongue, which include deformation for both single-region and multi-region contacts. No visual penetration between the tool and the tongue can be perceived. The interaction process is illustrated by the online video. During the deformation, the root and the bottom layer of the tongue were fixed.

In the experiments, computational stability of the graphic tool and force/torque stability are maintained during the manipulation. No abrupt jump or vibration of the graphic tool occurs between two adjacent simulation time steps. The curves of contact force/torque as functions of time during the tool–tongue interaction are also shown in Fig. 4.14.

To test the realism of tongue deformation by the method, the results are compared against the simulation of the same tongue model using the commercial software ANSYS,[2] which uses finite element analysis (FEA) (see Fig. 4.15). The parameters in the spring–sphere model can be tuned, including spring stiffness of the tongue and damping in order to achieve an acceptable simulation error between the results of the method and the FEA method by using ANSYS.

In Fig. 4.15, physical properties of the tongue were assigned the following values in the FEA model: Young's modulus: 100 kPa and Poisson's ratio: 0.49 (Payan et al. 1998). In the sphere model with springs, spring stiffness of the tongue and damping are set as 0.5 N/mm and 0.6 Ns/mm, respectively. As a boundary condition for the FEM simulation, the root and the bottom layer of the tongue were fixed. A static external load was exerted to three different locations on the tongue: from the top of the tongue's tip, from the lateral side, or from the top surface of the tongue in a distributed fashion.

---

[2] http://www.ansys.com/Products/Simulation+Technology/Systems+&+Multiphysics.

We can see that the shapes of the deformed tongue are similar from the two methods when we exert equal point loads in the two models. The geometric error of the corresponding deformed points on the two models is less than 5 %. As a further step, comparisons can be carried out to evaluate the simulation fidelity under dynamic loads.

4.8.2 Simulating Multi-region Hybrid Contacts

In a typical periodontal operation, the dentist manipulates a dental probe in the narrow periodontal pocket between a tooth and its surrounding gingiva. The method simulates the local deformation of the gingiva and the multi-region contacts between the probe and hybrid tissues (e.g., both the deformed gingiva and the rigid

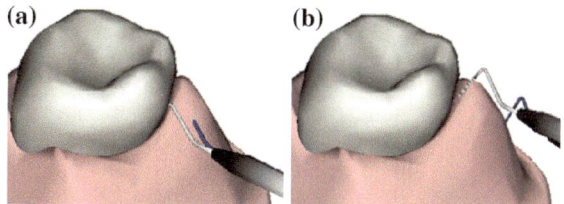

Fig. 4.16 No penetration between the graphic tool (*white*) and the tooth/gingiva. **a** The graphic tool contacts both the gingiva and the tooth in a hybrid contact. **b** The graphic tool only contacts the gingival. Copyright © IEEE. All rights reserved. Reprinted, with permission, from Wang et al. (2014). **a** Hybrid contacts and **b** unilateral contact

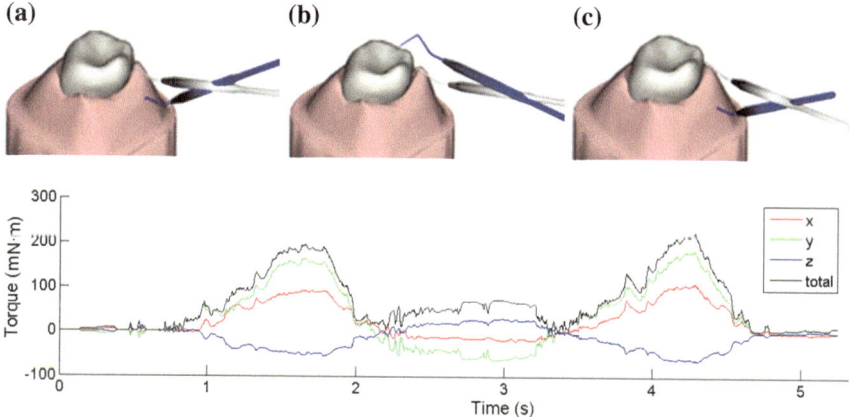

Fig. 4.17 A sequence of tool motion to show performance under contact switches: **a** Clockwise torque. **b** Counterclockwise torque. **c** Clockwise torque. Copyright © IEEE. All rights reserved. Reprinted, with permission, from Wang et al. (2014)

tooth) in such an operation. As shown in Fig. 4.16, non-penetration of the probe can be maintained before and after deformation. Small clearance of the periodontal pocket can be simulated, and there are no visual artifacts (i.e., no interpenetrations or separations between the graphic tool and the object).

Figure 4.17 shows snapshots of a sequence of tool motion and the corresponding torques. When the dentist slightly rotates the probe inside the periodontal pocket, contact switches occur because of the small clearance. From the torque curve, we can see that the torque signal reflects the change of the contact states responsively. The direction of the torque signal changes when the rotation direction of the probe changes in the sequence in Fig. 4.17. The force/torque feedback is stable during the whole manipulation sequence. A video showing the haptic interaction process is available online from the book Website http://extras.springer.com/.

In the motion sequence of insertion in Fig. 4.18, with the movement of the haptic tool, contact states change from unilateral contacts to bilateral contacts. Correspondingly, the simulated different force/torque components reflect the change of subtle force feelings under different contact states. When the tool is deliberately rotated under the bilateral contacts state (as shown in Fig. 4.18b, c), direction of torque signals changes responsively, and the simulation is stable. When the tool reaches the bottom of the pocket, an obvious force threshold can be felt, which provides helpful cues for the dentist to stop insertion along the vertical direction.

4.8.3 Bimanual Operation with Deformation

The method is computationally efficient. In Table 4.2, the time cost of each computation module during large deformation of the tongue is provided for two cases. In Case 1, level 2 of the sphere trees is used as the leaf nodes for collision detection and optimization, while in Case 2, level 3 of the sphere trees is used for collision detection and optimization. In both cases, the skeleton for deformation computation is defined in the level 2 of the sphere trees.

Furthermore, Fig. 4.19 shows the simulation of a bimanual operation within a complex oral cavity to test the performance of the method. Two Phantom Omnis are utilized to simulate the deformation of the tongue and that of the gingiva and thus to train the coordination between two hands' operations in a narrow space. See Table 4.3 for a summary of the data in this environment. The haptic interaction is stable. A video showing the interaction process is available online from the book Website http://extras.springer.com/.

4.8.4 Discussion

Theoretically, it is possible to improve the accuracy (i.e., of the stress, deformation, etc.) of the mass-spring model through an iterative comparison and parameter

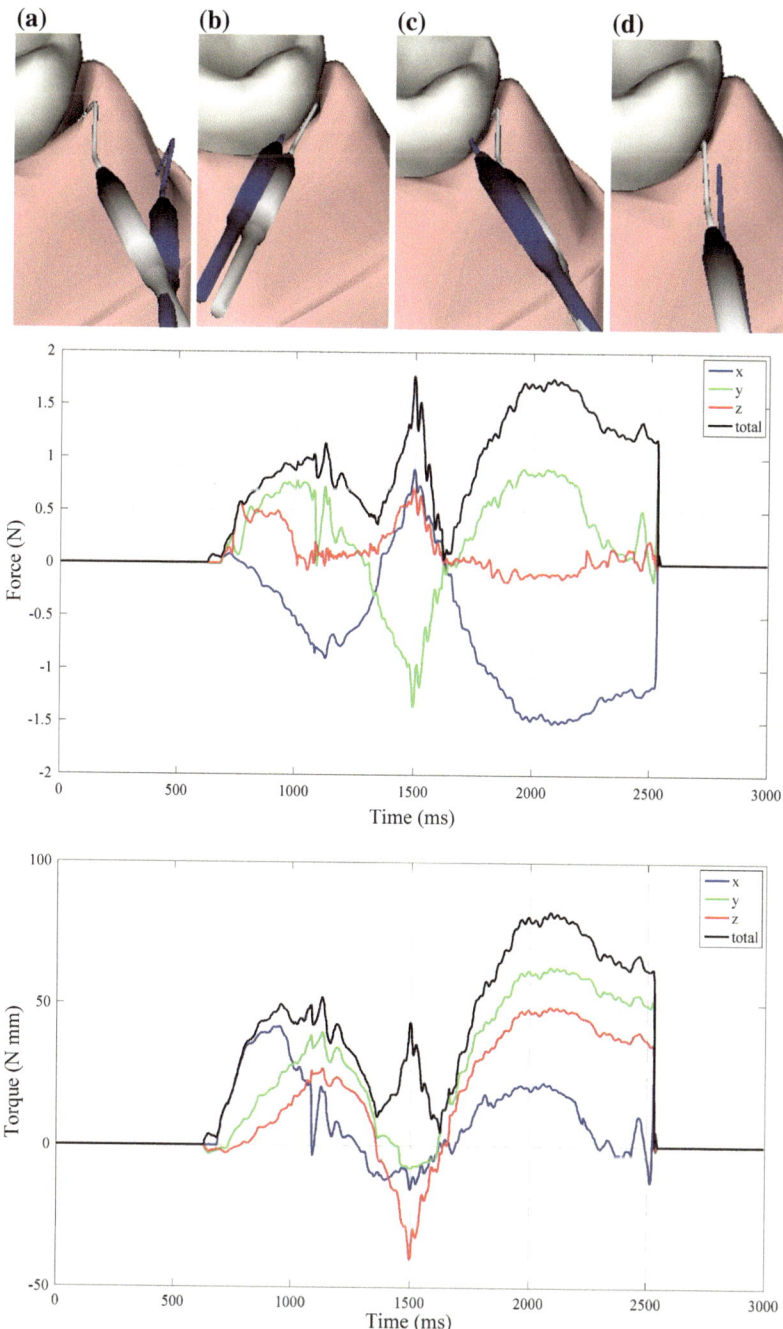

Fig. 4.18 A motion sequence of probe during pocket depth measurement: **a** unilateral resistance force from gingival. **b** Bilateral resistance forces from hybrid contacts. **c** Bilateral resistance forces from hybrid contacts and with deliberate tool rotation. **d** Resistance force from the pocket bottom. Copyright © IEEE. All rights reserved. Reprinted, with permission, from Wang et al. (2014)

Table 4.2 Time cost of each computation module

Two cases	Collision detection (ms)	Deformation (ms)	Optimization (ms)	Haptic loop (Hz)
Case 1	Ave: 0.13	Ave: 0.34	Ave: 0.16	Ave: 1587
	Max: 0.22	Max: 0.36	Max: 0.25	Min > 1205
Case 2	Ave: 0.20	Ave: 0.34	Ave: 0.13	Ave: 1492
	Max: 0.42	Max: 0.45	Max: 0.20	Min > 900

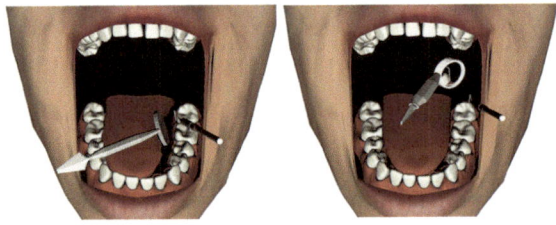

Fig. 4.19 Simulation of bimanual operation

Table 4.3 Data in the bimanual operation within a complex oral cavity. Copyright © IEEE. All rights reserved. Reprinted, with permission, from Wang et al. (2014)

Model	Polygons	Number of spheres
Lower jaw	4,000 × 14	512 × 14 teeth
Upper jaw	4,000 × 14	512 × 14 teeth
Gingiva of lower jaw	2,280 × 14	512 × 14 sub-gingiva
Gingiva of upper jaw	2,280 × 14	512 × 14 sub-gingiva
Tongue	1,045	512
Dental mirror	1,856	512
Dental probe	1,088	512

tuning process against a benchmark (e.g., a FEM model). In order to achieve this goal, the main technical challenge is to integrate an FEM solver into the simulation framework of the mass-spring model. In the dental simulation system, we only carried out preliminary tuning of the parameters of the mass-spring model through a manual process. The obtained error (5 %) is small enough for the training application, and therefore, we did not further improve the accuracy through a more rigorous parameterization process.

As a sphere tree is an approximation of a real shape, geometric error will lead to perceptible force/torque artifact when user slides along the surface. From the previous work (Wang et al. 2013), we have found that the force error (defined in Eq. 2.34) is dependent on the level of details of the sphere tree. For a curved surface (i.e., an ellipsoid model), the error is smaller than 10 % when a level-2 sphere tree is used and is smaller than 5 % when a level-3 sphere tree is used. In the simulation, a level-2 sphere tree with springs was used for the tongue deformation model, a level-3 sphere tree was used for the rigid tool, and a level-3 sphere tree was used for the gingiva. When the user slides the tool slowly along the surface of the gingiva, small bumps on the surface may be felt. When he/she slides fast, however, the surface of the tongue is felt smooth.

The geometric error of the method could be decreased by using more detailed sphere-tree models and using more accurate deformation models based on FEM. However, it is a challenge to ensure that the computation time is compatible with haptic rates. GPU is a possible solution to accelerate the computation process. In the next step, we plan to accelerate computation by using a GPU to support a deformation model of level 3 or level 4. While it is normally assumed that a tree structure is not easy to be parallelized, spatial partition with bounding boxes could be another solution to accelerate the collision detection process. All these issues need to be explored in the future study.

4.9 Summary

In this chapter, we have introduced a method for haptic simulation involving deformable objects in multi-region and hybrid contacts.

Using a sphere tree with springs to model a deformable object, the method efficiently computes large deformation (such as that of a tongue) as well as local deformation (such as that of gingivae) under contact. The method uses configuration-based optimization to compute the six-dimensional configuration of the graphic tool to avoid penetration of the tool into the deformed object even in multi-contact and hybrid contact scenarios and to compute the six-dimensional force/torque feedback stably without using virtual coupling.

The method of simulating hybrid contacts of the dental probe with both a rigid tooth and deformed gingivae is also suitable for other applications involving closely located hybrid contacts.

There are several possible directions for future research. One is to exploit computing power of GPU to deal with large-scale virtual environments with more spheres in deeper sphere trees. Another is to develop a continuous collision detection method to avoid pop through during interaction with thin objects.

One possible further study is to check whether the method of using a simple deformation model and executing all computations in a single loop at the high frequency of 1 kHz produces more realistic feeling or not comparing to existing work of using multi-threads for different computations with a more precise deformation model (such as one based on FEM). In the future, it is also worth extending the study to other surgical operations which require more delicate deformation, such as in cutting, suturing, and knotting.

References

Barbič J, James DL (2008) Six-DoF haptic rendering of contact between geometrically complex reduced deformable models. IEEE Trans Haptics 1(1):39–52

Bro-Nielsen M (1998) Finite element modeling in surgery simulation. Proc IEEE 86:490–503

Conti F, Khatib O, Baur C (2003) Interactive rendering of deformable objects based on a filling sphere modeling approach. In: Proceedings of the 2003 IEEE international conference on robotics and automation, Taipei, Taiwan, pp 3716–3721, 14–19 Sept 2003

Courtecuisse H, Jung H, Allard J, Duriez C, Lee DY, Cotin S (2010) GPU-based real-time soft tissue deformation with cutting and haptic feedback. Prog Biophys Mol Biol 103:159–168

Delingette H, Subsol G, Cotin S, Pignon J (1994) A craniofacial surgery simulation testbed. Visual Biomed Comput 2359(1):607–618

Duriez C, Dubois F, Kheddar A, Andriot C (2006) Realistic haptic rendering of interacting deformable objects in virtual environments. IEEE Trans Visualization Comput Graph 12 (1):36–47

Fierz B, Spillmann J, Harders M (2011) Element-wise mixed implicit-explicit integration for stable dynamic simulation of deformable objects. In: Proceedings of ACM/Eurographics symposium on computer animation 2011

Forsslund J, Sallnas E-L, Palmerius K-J (2009) A user-centered designed FOSS implementation of bone surgery simulations. In: World haptics conference. World haptics 2009, pp 391–392

Garre C, Otaduy MA (2010) Haptic rendering of objects with rigid and deformable parts. Comput Graph 34(6):689–697

Garre C, Hernández F, Gracia A, Otaduy MA (2011) Interactive simulation of a deformable hand for haptic rendering. In: The proceedings of the IEEE world haptics conference, Istanbul, Turkey, June 2011

Lin MC, Otaduy M (2008) Haptic rendering: foundations, algorithms, and applications. A K Peters, Ltd., Natick

Luciano C, Banerjee P, DeFanti T (2009) Haptics-based virtual reality periodontal training simulator. Virtual Reality 13(2):69–85

Luo Q, Xiao J (2007) Contact and deformation modeling for interactive environments. IEEE Trans Rob 23(3):416–430

Mendoza C, O'Sullivan C (2005) An interruptible algorithm for collision detection between deformable objects. In: Workshop on virtual reality interaction and physical simulation (2005), The Eurographics Association

Meier U, Lopez O, Monserrat C, Juan M, Alcaiz M (2005) Real-time deformable models for surgery simulation: a survey. Comput Methods Programs Biomed 77(3):183–197

Müller M, Heidelberger B, Teschner M, Gross M (2005) Meshless deformations based on shape matching. ACM Trans Graph 24(3):471–478

Nealen A, Muller M, Keiser R, Boxerman E, Carlson M (2006) Physically based deformable models in computer graphics. Comput Graph Forum 25(4):809–836

Payan Y, Bettega G, Raphael B (1998) A biomechanical model of the human tongue and its clinical implications. In: Medical image computing and computer-assisted intervention, MICCAI'98. Lecture notes in computer science, vol 1496/1998, pp 688–695

Peterlik I, Nouicer M, Duriez C, Cotin S, Kheddar A (2011) Constraint-based haptic rendering of multirate compliant mechanisms. IEEE Trans Haptics 4:175–187 (special issue on haptics in medicine and clinical skill acquisition)

Saupin G, Duriez C, Cotin S (2008) Contact model for haptic medical simulations, ISBMS 2008, pp 157–165

Shi W, Payandeh S (2010) Towards point-based haptic interactions with deformable objects. In: ASME 2010 world conference on innovative virtual reality (WINVR2010), Ames, Iowa, USA, pp 259–265, 12–14 May 2010

Tse B, Harwin W, Barrow A, Quinn B, San Diego J, Cox M (2010) Design and development of a haptic dental training system—hapTEL. In: EuroHaptics 2010 conference. Lecture notes in computer science, vol 6192/2010. VU University, Amsterdam, pp 101–108

Wang D, Zhang Y, Wang Y, Lü P, Zhou R, Zhou W (2009) Haptic rendering for dental training system. Sci China Ser F Inf Sci 52(3):529–546

Wang D, Zhang X, Zhang Y, Xiao J (2013) Configuration-based optimization for six degree-of-freedom haptic rendering for fine manipulation. IEEE Trans Haptics 6(2):167–180

Wang D, Shi Y, Liu S, Zhang Y, Xiao J (2014) Haptic simulation of organ deformation and hybrid contacts in dental operations. IEEE Trans Haptics 7(1):48–60

Wu X, Downes M, Goktekin T, Tendick F (2001) Adaptive nonlinear finite elements for deformable body simulation using dynamic progressive meshes. Comput Graph Forum 20(3):349–358

Zhong Y, Shirinzadeh B, Alici G, Smith J (2005) A new methodology for deformable object simulation. In: Proceedings of 2005 IEEE international conference on robotics and automation, pp 1902–1907

Zhuang Y, Canny J (2000) Haptic interaction with global deformations. In: The international conference on robotics and automations, IEEE, San Francisco, California, pp 2428–2433, 24–28 April 2000

Chapter 5
Evaluation of Haptic Rendering Methods

Performance evaluation of haptic rendering algorithms is an important topic for measuring the fidelity of haptic simulation. In this chapter, we first briefly survey the progress in this field and then introduce a measurement-based method to evaluate the accuracy of different 6-DoF haptic rendering algorithms in interactions involving multi-region contacts. A measurement system is constructed to measure 6-D force/torque between a hand-held tool and a target object, and a four-camera motion tracking system is used to measure real-time position and orientation of the tool. A computational model of the haptic tool's location is proposed based on the measured force/torque and motion signal of the hand-held tool. The effectiveness of the evaluation method is illustrated by computing the force/torque error of two haptic rendering methods: an adapted Voxmap pointshell method and a configuration-based optimization method.

5.1 Related Literature

The performance of haptic simulation is determined by both haptic devices and haptic rendering algorithms. There is an evaluation of haptic devices in the literature (Samur 2010; Salisbury et al. 2011). Similarly, evaluation of haptic rendering methods has been widely studied, which is necessary to evaluate how haptic rendering algorithms can achieve high realism and accuracy in tasks such as surgical simulation.

In related literature, there are three kinds of methods for evaluating haptic rendering algorithms: perception-based (Luciano et al. 2009; Steinberg et al. 2007; Johnson et al. 2000), theoretical model-based (Weller et al. 2010; Sagardia et al. 2009), and measurement-based (Ruffaldi et al. 2006; Pai et al. 2001; Kuchenbecker et al. 2009; Hoever et al. 2008) methods.

Perception-based evaluation was carried out based on a user's subjective feeling, and statistical scores can be computed for different rendering algorithms. Luciano et al. (2009) developed a dental training prototype called PerioSim. Typical periodontal procedures were used as target tasks. Several dentists were invited to use

© Springer-Verlag Berlin Heidelberg 2014
D. Wang et al., *Haptic Rendering for Simulation of Fine Manipulation*,
DOI 10.1007/978-3-662-44949-3_5

the system and provide evaluation according to a seven-leveled score sheet (Steinberg et al. 2007). Similarly, Johnson et al. (2000) carried out subjective evaluation on decay simulation. However, these methods provide only qualified evaluation and not quantified comparison among different algorithms. Leskovsky et al. (2006) used multidimensional scaling to assess the fidelity of haptic interactions, and they analyzed similarity ratings provided by users comparing pairs of haptically presented objects. The authors mentioned that a large number of participants were required to get more rigorous results.

Weller et al. (2010) proposed a theoretical model-based approach to evaluate the fidelity of collision response methods for haptic rendering. Several objects of simple shapes, such as sphere and cylinder, were selected as benchmarks. Given the trajectory of the haptic tool, reference contact force/torque signals were computed by using a theoretical collision response model. The fidelity criterion was defined as the error between the simulated force signal and reference force signal. While this method can provide quantitative evaluation scores, it is only suitable for interactions between simple-shaped objects because the theoretical collision response model could not solve the multi-region contacts between a pair of objects with complex shapes.

In order to evaluate the interaction force involving objects of complex shapes, such as a Stanford bunny or a dragon, measurement-based evaluation methods for 3-DoF haptic rendering were widely studied. Ruffaldi et al. (2006) proposed a standardized evaluation method of haptic rendering systems. They used a force sensor to measure the contact force between the probe and an object, and used a six-camera motion-capture system to measure the position of the probe. After getting the physical contact force and the moving trajectory of the probe's tip (which is called Out-trajectory), they computed an In-trajectory by offsetting the Out-trajectory along the inverted normal direction of the contact normal force. This In-trajectory is then input into the target haptic rendering algorithms to simulate the trajectory of the hand-held haptic device. Similarly, Pai et al. (2001) proposed a system to measure the deformation, contact sound, and contact force between the probe and measured objects. Kuchenbecker et al. (2009) proposed a measurement system to measure the acceleration and vibration signals. Hoever et al. (2008) proposed a measurement system to measure the contact force and damping coefficient between the probe and various viscoelastic objects. In these methods, a mechanical arm such as a Phantom haptic device was usually used to track the position/orientation of the hand-held tool.

There is no quantified method for evaluating complex haptic interactions involving multi-region contacts. Existing theoretical model-based evaluation can only deal with simple-shaped objects. The perception-based methods can evaluate interactions involving any kind of objects but can only provide subjective and qualitative evaluation results. Existing measurement-based methods were only focused on 3-DoF rendering tasks (Ruffaldi et al. 2006; Pai et al. 2001) involving point contacts between a tool and an object.

5.2 Objective Evaluation Based on Measurement Data

In this section, we introduce a measurement-based method to evaluate the accuracy of different 6-DoF haptic rendering algorithms for interactions involving multi-region contacts. Section 5.2.1 outlines the framework of the method. Section 5.2.2 proposes quantified metrics for accuracy. Section 5.2.3 introduces the measurement equipment. Section 5.2.4 introduces a real-time measurement and computation method of force/torque and pose signals. Section 5.2.5 reviews two 6-DoF haptic rendering algorithms as examples for testing the evaluation method. Finally, Sect. 5.2.6 provides evaluation results of the two target algorithms.

5.2.1 Framework

The framework of a generic measurement-based evaluation method is illustrated in Fig. 5.1. Three types of tools are involved: a *hand-held tool* for real-world experiment, the corresponding *haptic tool* for virtual simulation of the real-world experiment that is specific to a haptic device, and the *graphic tool* that is the graphical display of the haptic tool during virtual contact. The essence of the approach is to compare measured configuration and force–torque signals of the hand-held device in real-world interaction with a target object with those rendered from the corresponding haptic interaction in the virtual world that simulates the real-world experiment. The real-world measurements form the ground truth to evaluate the accuracy of haptic rendering algorithms. To achieve that, we address three key issues:

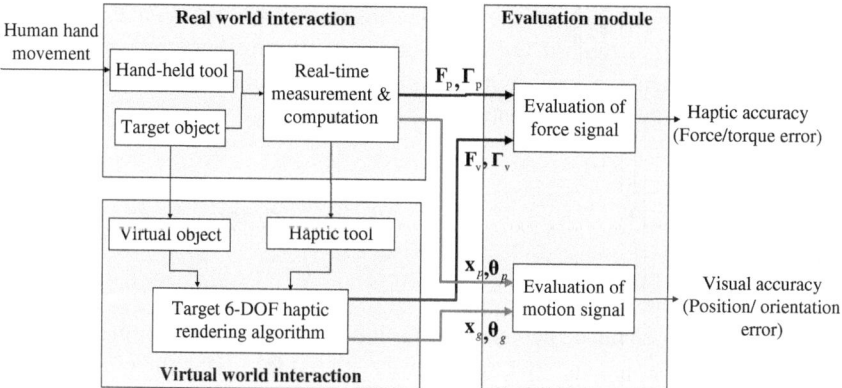

Fig. 5.1 Framework of a measurement-based evaluation method

(a) how to define the metrics of accuracy relating the real-world measurements to the virtual-world simulated results (Sect. 5.2.2);
(b) how to measure the trajectory (i.e., configuration-time sequence) and the force/torque signals of the hand-held device during real-world interaction with a target object (Sect. 5.2.4);
(c) how to determine the corresponding trajectory (i.e., configuration-time sequence) of the haptic tool from the measured trajectory of the hand-held tool during the real-world experiment while taking into account the limited stiffness of a haptic device, which often leads to the haptic tool penetrating into a virtual object during interaction (Sect. 5.2.4).

5.2.2 Metrics of Accuracy

We define two types of accuracies for evaluation:

1. *Haptic accuracy*: measures the root-mean-square error between the force/torque produced in the real world when a tool interacts with an object and the simulated force/torque between the corresponding haptic tool and the virtual object;

$$\sigma_F = \sqrt{\sum_{i=1}^{M} \frac{\|\mathbf{F}_p - \mathbf{F}_v\|^2}{M}}, \quad \sigma_T = \sqrt{\sum_{i=1}^{M} \frac{\|\mathbf{\Gamma}_p - \mathbf{\Gamma}_v\|^2}{M}} \tag{5.1}$$

where $\mathbf{F}_p$ and $\mathbf{F}_v$ are the three-dimensional force signal from the real and virtual world, respectively. $\mathbf{\Gamma}_p$ and $\mathbf{\Gamma}_v$ are the three-dimensional torque signal from the real and virtual world, respectively. M is the number of sampling points during the measurement process in the real world.

2. *Visual accuracy*: measures the root-mean-square error between the position/orientation of the tool in the real world and the position/orientation of the graphic tool, which should minimize virtual penetration or separation when it interacts with the virtual object:

$$\sigma_x = \sqrt{\sum_{i=1}^{M} \frac{\|\mathbf{x}_p - \mathbf{x}_g\|^2}{M}}, \quad \sigma_\theta = \sqrt{\sum_{i=1}^{M} \frac{\|\mathbf{\theta}_p - \mathbf{\theta}_g\|^2}{M}} \tag{5.2}$$

where $\mathbf{x}_p$ and $\mathbf{x}_g$ indicate the three-dimensional position signals of the real hand-held tool and the graphic tool, respectively. $\mathbf{\theta}_p$ and $\mathbf{\theta}_g$ are the three-dimensional orientation signals of the real hand-held tool and the graphic tool, respectively.

The smaller the mean squared error, the more accurate the haptic rendering algorithm is in reflecting the real-world interaction.

Based on the above metrics, both strength and limitation of a target haptic rendering algorithm can be identified. The findings can be used for identifying possible bottleneck and ways to improve the performance of target haptic rendering algorithms.

5.2.3 Measurement System for Force and Motion Capture

An example of hand-held measurement equipment is shown in Fig. 5.2. A six-axis force sensor is used to measure three-dimensional force/torque between the hand-held probe and the target object (Wang et al. 2011). A four-camera motion tracking system is used to measure real-time position and orientation of the probe (Fig. 5.3).

During the measurement process, a human operator manipulates the probe of the equipment to contact the surface of the target object. The equipment is used to capture two kinds of signals: (1) the contact force and torque between the probe and the object and (2) the real-time position and orientation of the probe.

Fig. 5.2 Operator manipulates the probe to contact the object

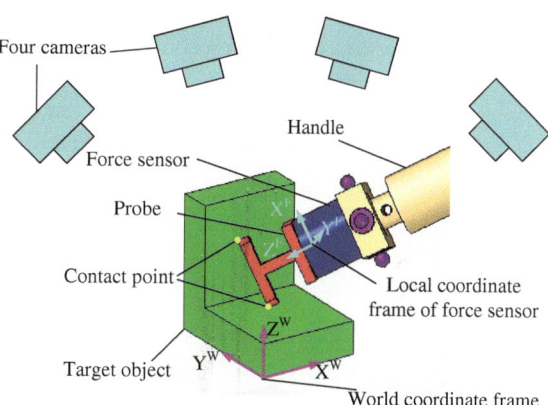

Fig. 5.3 Overview of the hardware platform

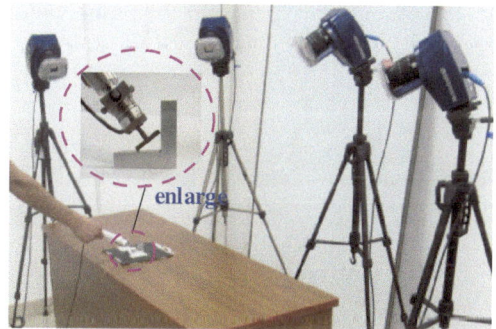

5.2.4 Computational Model of the Haptic Tool's Location

Based on the measured force/torque and position/orientation signal on a hand-held tool, we can formulate and compute the position and orientation of the haptic tool in a systematic way and use the results as a common sequence of input signals for various target haptic rendering algorithms.

5.2.4.1 Measurement of Force and Torque of the Hand-held Tool

Real-time measurement and computation for the hand-held tool include the following steps:

1. Compute the position and orientation of the hand-held tool with respect to the world frame:

    (a) measure the real-time motion of the dynamic markers on the tool;
    (b) compute the position and orientation of the hand-held tool with respect to the world frame;

2. Compute the force/torque with respect to the world frame:

    (a) measure the contact force/torque with respect to the local frame of the force/torque sensor;
    (b) compute the contact force/torque at the center of the hand-held tool with respect to the coordinate frame of the baseboard.

By measuring the orientation of the hand-held probe, we establish the transformation from the local frame of the force sensor to the normal of a contact point to compute the contact force along the normal direction. The force with respect to the world coordinate frame is

$$\mathbf{F}^{W} = \mathbf{R}_{F}^{W} \cdot \mathbf{F}^{F} \tag{5.3}$$

where $\mathbf{F}^{F}$ denotes the measured force vector in the local frame of the force sensor, $\mathbf{F}^{W}$ denotes the measured force in the world frame. $\mathbf{R}_{F}^{W}$ denotes the rotational transformation matrix from the frame of the force sensor to the world frame.

The torque at the center of the probe with respect to the world frame is

$$\boldsymbol{\Gamma}^{W} = \boldsymbol{\Gamma}^{A} + \mathbf{R}_{F}^{W} \cdot \boldsymbol{\Gamma}^{F} \tag{5.4}$$

where $\boldsymbol{\Gamma}^{F}$ denotes the measured torque signal with respect to the local frame of the force/torque sensor and $\boldsymbol{\Gamma}^{A}$ denotes an additional torque, which is derived as

$$\boldsymbol{\Gamma}^{A} = \mathbf{r} \times \mathbf{F}^{W} \tag{5.5}$$

where $\mathbf{r}$ denotes the vector from the center position of the tool to the center of the force/torque sensor with respect to the world frame as shown in Fig. 5.2.

5.2.4.2   Computation of Position/Orientation of the Haptic Tool

Here, the issue is how to determine the position and orientation of the haptic tool in the virtual environment corresponding to the measured pose and associated force/torque signals of a hand-held tool. When the hand-held tool contacts with a rigid target object in the real world, there is no penetration because the stiffness of the real object is relatively high. However, since a haptic device has only limited stiffness, applying the same force/torque that the real-world hand-held tool is applied to the haptic tool will result in the haptic tool penetrating into the virtual object. Thus, the pose of the haptic tool has to be different from the measured pose of the hand-held tool. We take into account the limited stiffness of a haptic device by using a spring model to compute the position of the haptic tool (which will penetrate into the virtual object's surface when contacts occur) as follows

$$\mathbf{x}_h^W = \mathbf{x}_p^W - \mathbf{F}^W / k_S \tag{5.6}$$

where $\mathbf{x}_p^W$ denotes the position of the hand-held tool in the real world. k_S denotes the translational stiffness of the target haptic device (e.g., 1 N/mm for the Phantom Premium device).

For computing the rotation matrix of the haptic tool, we assume that there exist a virtual three-dimensional torsional spring between the hand-held tool and the haptic tool. When contacts occur between the hand-held tool and the target object, a contact torque will be produced. Then, the orientation of the haptic tool will deviate from the orientation of the hand-held tool. The rotation matrix of the haptic tool with respect to the world frame can be computed as

$$\mathbf{R}_h^W = \mathbf{R}_p^W \cdot \mathbf{R}_h^p \tag{5.7}$$

where $\mathbf{R}_p^W$ denotes the rotational matrix of the hand-held tool with respect to the world frame. $\mathbf{R}_h^p$ denotes the rotation matrix representing the orientation difference between the hand-held tool and the haptic tool, which can be derived from an equivalent rotation angle θ_h^p about the corresponding rotational axis vector $\mathbf{k}_h^p$ as

$$\begin{cases} \theta_h^p = \|\mathbf{\Gamma}^F\| / k_\theta \\ k_h^p = -\mathbf{\Gamma}^F / \|\mathbf{\Gamma}^F\| \end{cases} \tag{5.8}$$

where $\mathbf{\Gamma}^F$ denotes the measured torque signal, and k_θ denotes the rotational stiffness of the target haptic device. Using the equivalent angle-axis method (Craig 1989), we can easily derive

$$\mathbf{R}_h^p = \begin{pmatrix} k_x k_x v\theta_h^p + c\theta_h^p & k_x k_y v\theta_h^p - k_z s\theta_h^p & k_x k_z v\theta_h^p + k_y s\theta_h^p \\ k_x k_y v\theta_h^p + k_z s\theta_h^p & k_y k_y v\theta_h^p + c\theta_h^p & k_y k_z v\theta_h^p - k_x s\theta_h^p \\ k_x k_z v\theta_h^p - k_y s\theta_h^p & k_y k_z v\theta_h^p + k_x s\theta_h^p & k_z k_z v\theta_h^p + c\theta_h^p \end{pmatrix} \qquad (5.9)$$

where

$$\begin{cases} \mathbf{k}_h^p = \begin{bmatrix} k_x, k_y, k_z \end{bmatrix}^T \\ s\theta_h^p = \sin\theta_h^p \, , \, c\theta_h^p = \cos\theta_h^p \, , \, v\theta_h^p = (1 - \cos\theta_h^p) \end{cases} \qquad (5.10)$$

The homogeneous transformation matrix of the position/orientation of the haptic tool with respect to the world frame is

$$\mathbf{T}_h^W = \begin{pmatrix} \mathbf{R}_h^W & \mathbf{x}_h^W \\ 0_{1\times3} & 1 \end{pmatrix} \qquad (5.11)$$

This matrix represents the input signal to target haptic rendering algorithms, as will be explained in Sect. 5.2.5.

5.2.5 Two Example Rendering Methods

Existing haptic rendering algorithms can be classified based on the collision response methods used and the geometric models for the tool and the manipulated objects (Otaduy and Lin 2006; Ortega et al. 2007; Johnson et al. 2005). We choose two typical haptic rendering methods as the examples for evaluation using the measurement-based approach: a penalty-based method adapted from the VPS method (Barbič and James 2008) and a constraint-based method using sphere trees (Wang et al. 2013), as described in Chap. 2. The measurement-based evaluation approach is general and can be used to evaluate other haptic rendering algorithms.

In the VPS method, a point-based representation is used for one object, and a signed-distance field is used for the other object. This model is related to the original Voxmap pointshell method (McNeely et al. 2006). The collision response is computed by a penalty-based approach, i.e., the motion of the graphic tool is computed by integrating static equilibrium equations based on the Newtonian principle. In order to produce stable haptic feedback, static virtual coupling combined with quasi-static damping is adopted in this method. We downloaded codes from http://www-bcf.usc.edu/~jbarbic/code/index.html to implement the adapted version of the VPS method.

The configuration-based 6-DoF haptic rendering algorithm is described in detail in Chap. 2 for simulating contacts between a rigid tool and another rigid object. Both the tool and the object are represented by sphere trees. The position and orientation of the graphic tool (as the avatar of the haptic tool) in the virtual environment are constrained by contacts and determined by solving a constrained optimization problem.

5.2.6 Experimental Results

Experiments were carried out to validate the measurement-based evaluation approach by evaluating the two example rendering algorithms. For both rendering methods, the translational stiffness was set as 0.5 N/mm, and the rotational stiffness was set as 500 m Nm/rad.

Because the sampling rate of the sensing system in the real environment is 100 Hz, which is lower than the typical haptic rendering rate (i.e., 1 kHz), one solution is to interpolate between consecutive position/orientation values of the haptic tool computed by Eq. (5.11) to obtain more poses in order to have pose samples of the haptic tool at 1 kHz. Using the new pose-time signals as an input to a haptic rendering algorithm, virtual force/torque at 1 kHz could be computed. In the experiments reported in this chapter, because the moving velocity of the hand-held tool was very low, there was hardly any change between poses of the haptic tool computed from two consecutively measured poses of the hand-held tool with force/torque values. Therefore, no interpolation was conducted.

Figure 5.4 shows two benchmark tests: One was to form a single-region contact between a tool and a flat surface; the other was to form a multi-region contact between a tool and an L-shaped object. A Nano 17 by ATI Inc. is used as the force sensor to provide six-dimensional force/torque measurement (with a high resolution of 1/320 N). A four-camera motion tracking system produced by Motion Analysis Inc. is used to measure real-time position and orientation of the probe (with nominal resolution of around 0.1 mm for each marker). In order to alleviate the effect of friction, we used special materials (brass for the object and aluminum for the probe, thus the friction coefficient was 0.02) and good lubrication to minimize friction.

We measured the accuracy of each rendering algorithm in the experiments. The force/torque signals during a single-region contact and a multi-region contact are shown in Figs. 5.5 and 5.6, respectively. The error data are shown in Table 5.1.

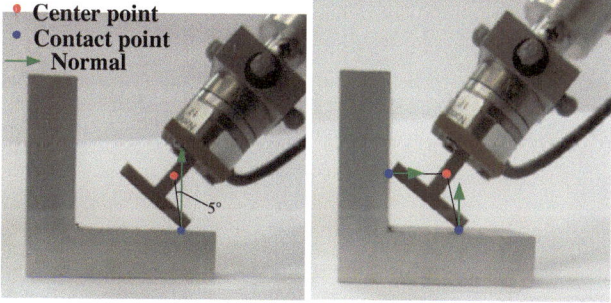

Fig. 5.4 Two contact tests: *left* single region; *right* multi-region

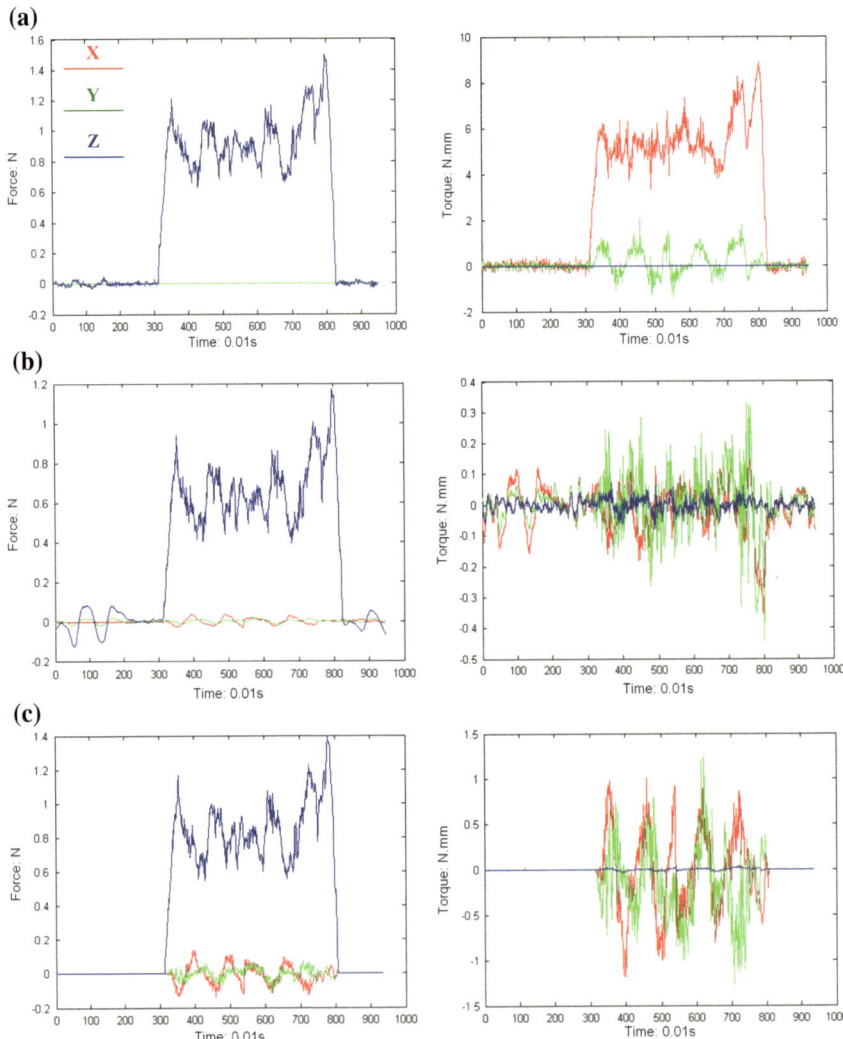

Fig. 5.5 Contact force/torque signals from a single-region contact. **a** Measured force/torque signals, **b** contact force and torque modeled by the adapted VPS method (160 × 160 × 160 voxels/ 98,712 points), **c** contact force and torque modeled by the constraint-based method (32,768 spheres)

For the adapted VPS method, models with different levels of detail were used to compare their performances. For the tool, we used 80 × 80 × 80 voxels or 160 × 160 × 160 voxels. For the static object, we used 2,438/24,365/39,322/98,712 points on the point-shell model.

For the constraint-based method, models with two different levels of detail were used to compare their performances, i.e., 4,096 spheres and 32,768 spheres.

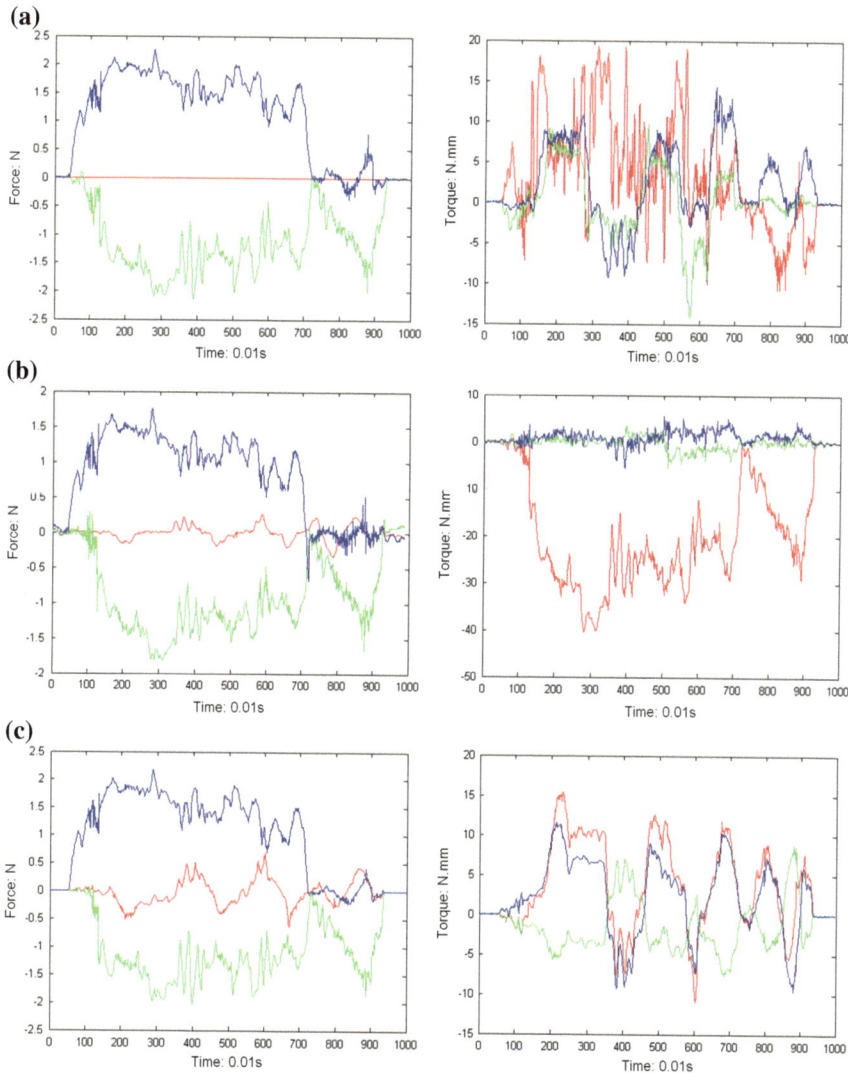

Fig. 5.6 Contact force/torque signals from multi-region contacts. **a** Measured force/torque signals, **b** contact force and torque modeled by the adapted VPS method (160 × 160 × 160 voxel/98,712 points), **c** contact force and torque modeled by the constraint-based method (32,768 spheres)

From the experimental results in Table 5.1, we observe that the force error of the adapted VPS method, when the highest resolution of 160^3 voxels/98,712 points was used, is greater than that of the constraint-based method, when the highest resolution of 32,768 spheres was used. This seems that the direction and magnitude of

Table 5.1 Force/torque errors for the two example haptic rendering methods

Example methods		Force (N)/Torque (N mm) error	
		Single region	Multi-region
The VPS method	80^3 voxel/2,438 points	0.65/4.42	1.53/12.42
	160^3 voxel/24,365 points	0.323/4.71	0.66/24.16
	160^3 voxel/39,322 points	0.297/4.46	0.62/25.12
	160^3 voxel/98,712 points	0.22/4.4	0.466/28.05
Optimization method	4,096 spheres	0.099/4.27	0.307/10.45
	32,768 spheres	0.085/4.09	0.276/9.74

the force was modified by virtual coupling in the adapted VPS method. When the tool slides across the sharp edge of the object surface, the force error was apparent. Furthermore, visual penetration occurred in the adapted VPS method, especially in the case of multi-region contacts.

For both methods, force/torque errors became smaller when finer details of geometric elements were used. However, the torque feedback errors were fairly large even when a very detailed object model was used. For the sphere tree-based optimization method, one possible reason was that the fluctuation of the feedback torque signal was caused by the rugged surface of adjacent spheres. The reason for the torque error of the adapted VPS method was not so clear and worth to be investigated.

We have provided the measured position/orientation signal of the hand-held tool, the measured force/torque signal and the computed position/orientation signal of the haptic tool on the Webpage (http://haptic.buaa.edu.cn/download/HapticEvaluationData.htm). Furthermore, the 3D geometric model of the tool and the L-shaped object are also provided. These data can be used by others to evaluate 6-DoF haptic rendering approaches using the proposed method.

The *visual accuracy* can be computed similar to the *haptic accuracy*. The pose signal of the graphic tool can be recorded during haptic rendering process, and the *visual accuracy* can be computed using Eq. (5.2).

Based on the mathematical model relating the errors of position/orientation and force/torque, if we expect the force/torque error to be smaller, the measured error of the markers' position should be smaller. This can be realized by using high-end motion-capture systems with higher accuracy. Although the nominal resolution of the adopted camera system is around 0.1 mm for each marker, the experiments' results showed that the distance between each pair of markers on the hand-held tool was fluctuating, and the maximum measurement error was about 0.35 mm. The possible reason is that the four-camera system is not sufficient to provide accurate measurement because occlusion of some markers can occur when the tool assumes certain postures. Obviously, accuracy of the method can be enhanced by using cameras of higher accuracy, which is a future research topic.

5.3 Summary

In this chapter, a measurement-based approach is introduced to enable quantified performance evaluation of 6-DoF haptic rendering algorithms considering both the haptic and visual accuracy.

A method is introduced to provide consistent pose signals of the haptic tool for evaluating target algorithms, which are computed based on measured signal of position/orientation signals and the force/torque signals of the hand-held tool during real-world interaction. Using such consistent input signals, the method avoids the subjective feeling of a human operator and the effects of any specific haptic display device.

A penalty-based method adapted from the VPS method and the configuration-based optimization method using sphere trees were used as two target algorithms to illustrate the implementation process of the evaluation method. An experiment with a tool interacting with an L-shaped object was carried out to validate the performance of the method. Experimental results show that the method can capture the real-time force/torque and position/orientation signals during multi-region contacts and can be utilized to quantitatively evaluate accuracies of typical 6-DoF haptic rendering algorithms. A preliminary database for public evaluation of 6-DoF haptic rendering algorithms is constructed for sharing in the haptics community.

References

Barbič J, James DL (2008) Six-dof haptic rendering of contact between geometrically complex reduced deformable models. IEEE Trans on Haptics 1(1):39–52

Craig JJ (1989) Introduction to robotics: mechanics and control, 2nd edn. Addison-Wesley, Reading

Hoever R, Harders M, Szekely G (2008) Data-driven haptic rendering of visco-elastic effects. In: Symposium on haptic interfaces for virtual environments and teleoperator systems, pp 201–208

Johnson L, Thomas G, Dow S, Stanford C (2000) An initial evaluation of the iowa dental surgical simulator. J Dent Educ 64(12):847–853

Johnson DE, Willemsen P, Cohen E (2005) Six degree-of-freedom haptic rendering using spatialized normal cone search. IEEE Trans Visual Comput 11(6):661–670

Kuchenbecker KJ et al (2009) Haptography capturing and recreating the rich feel of real surfaces. In: International symposium on robotics research, Aug 2009

Leskovsky P, Cooke T, Ernst M, Harders M (2006) Using multidimensional scaling to quantify the fidelity of haptic rendering of deformable objects. In: Proceedings of eurohaptics international conference (EuroHaptics'06), pp 289–296

Luciano C, Banerjee P, DeFanti T (2009) Haptics-based virtual reality periodontal training simulator. Springer Virtual Reality vol 13, no 2

McNeely W, Puterbaugh K, Troy J (2006) Voxel-based 6-dof haptic rendering improvements. Haptics-e vol 3, no 7

Ortega M, Redon S, Coquillart S (2007) A six degree-of-freedom god-object method for haptic display of rigid bodies with surface properties. IEEE Trans Visual Comput Graphics 13 (3):458–469

Otaduy MA, Lin MC (2006) A modular haptic rendering algorithm for stable and transparent 6-DoF manipulation. IEEE Trans Rob 22(4):751–762

Pai DK, Doel K, James DL et al (2001) Scanning physical interaction behavior of 3D objects. (*SIGGRAPH*), pp 87–96

Ruffaldi E, Morris D, Edmunds T et al (2006) Standardized evaluation of haptic rendering systems. In: IEEE Haptic Interfaces for Virtual Environment and Teleoperator Systems, *VR*

Sagardia M, Hulin T, Preusche C et al (2009) A benchmark of force quality in haptic rendering. In: International conference on human-computer interaction (HCI), San Diego, USA, 19–24 July 2009

Salisbury C, Gillespie RB, Tan HZ, Barbagli F, Salisbury JK (2011) What you can't feel won't hurt you: evaluating haptic hardware using a haptic sensitivity contrast function. IEEE Trans Haptics 4(2):134–146

Samur E (2010) Systematic evaluation methodology and performance metrics for haptic interfaces. PhD thesis, Ecole Polytechnique Federale de Lausanne (EPFL), Lausanne

Steinberg D, Bashook PG, Drummond J, Ashrafi S, Zefran M (2007) Assessment of faculty perception of content validity of periosim(c), A haptic-3D virtual reality dental training simulator. Dent Educ. 71(12):1574–1582

Wang Z, Wang D, Song T (2011) Spring-based measurement equipment for evaluation of 3-DOF haptic rendering algorithms. In: 10th IEEE international symposium on haptic audio visual environments and games, HAVE, Qinhuangdao, Hebei, China, 14–17 Oct 2011

Wang D, Zhang X, Zhang Y, Xiao J (2013) Configuration-based optimization for six degree-of-freedom haptic rendering for fine manipulation. IEEE Trans on Haptics 6(2):167–180

Weller R, Mainzer D, Zachmann G et al (2010) A benchmarking suite for 6-DOF real time collision response algorithms. In: Proceedings of the 17th ACM symposium on virtual reality software and technology 2010 (VRST'2010). Hong Kong, 2010

Chapter 6
Application: A Dental Simulator

In this chapter, we introduce a haptics-enabled dental simulator, iDental, as an application of the configuration-based optimization approach. Valuable lessons have been identified from the user evaluation results. Based on a brief survey of existing dental simulation systems, we introduce the function requirements and the features of the iDental system in Sect. 6.2. In Sect. 6.3, we discuss the modeling method for various oral tissues and pathological changes. In Sect. 6.4, two bimanual periodontal operation tasks are introduced to test the capability of the configuration-based optimization approach. A hybrid evaluation approach combining subjective and objective evaluation methods is proposed, and corresponding user study results are explained in Sects. 6.5 and 6.6. In Sect. 6.7, needed improvements for the hardware and software system are identified based on the results of user study.

6.1 Background

Although surgical simulators are yet to become a practical way for surgeon training (Aggarwal and Darzi 2009), surgical simulation has great potential because of both technical advantages and societal demands for training more qualified surgeons (Coles et al. 2010; Basdogan et al. 2007). An active area of research is simulation of dental operations in recent years. As sensorimotor skills are difficult to train using visual-only computer simulations, haptically enhanced simulation provides a promising alternative to enable sensorimotor involvement needed for dental training.

Unlike traditional training methods such as a Phantom head, a virtual reality dental simulation system can simulate both common and uncommon pathological changes, which may help dental students to learn diagnosis/operation skills. Furthermore, the three-dimensional motion of the surgical instruments and interaction force signals can be recorded and played back for analysis to benefit teaching and certification.

Electronic supplementary material The online version of this article (doi:10.1007/978-3-662-44949-3_6) contains supplementary material, which is available to authorized users.

D. Wang et al., *Haptic Rendering for Simulation of Fine Manipulation*,
DOI 10.1007/978-3-662-44949-3_6

Several dental simulators have been developed in academic or commercial institutions. In 1990s, Ranta and Aviles (1999) introduced the concept design of a virtual reality dental training (VRDT) system for practicing cavity preparation. Thomas et al. (2000) developed a training system with Impulse2000, enabling the operator to practice the detection of carious lesions. In recent years, more powerful dental simulators have been developed. PerioSim was developed for periodontal simulation, which can simulate three typical operations including pocket probing, calculus detection, and calculus removal (Luciano et al. 2009). This system is focused on probing the differences of different tissues around a tooth. In hapTEL, two generations of prototypes were developed based on feedback from user evaluation (Tse et al. 2010). Several companies have been focused on developing commercial dental training systems. Simodont was developed by MOOG Inc. and could simulate drilling and mirror reflection. Forsslund Dental system was developed to practice dental drilling and wisdom teeth extraction (Forsslund et al. 2009).

In existing dental simulators as shown in Table 6.1, only a few provide deformation simulation of oral organs. The Simodont system considered small, local deformation of the tongue. Large deformation of the tongue (for exposing the target teeth) was not addressed. None of the systems can simulate multiple contacts and corresponding force/torque feedback between the dental probe and both a rigid tooth and a deformed gingiva, which are common when the operator inserts a dental probe into the narrow periodontal pocket.

Performance evaluation is an important topic in surgical and dental simulation research (Quinn et al. 2007; Jasinevicius et al. 2004). Minimally invasive surgery trainer (MIST-VR) simulated typical laparoscopic skills such as suturing and diathermy (Crossan 2003). Kothari compared this system against Yale Laparoscopic Skills Course and validated the effectiveness of the simulator (Kothari et al. 2002). In PerioSim, statistical analysis based on scores from qualitative evaluation experiment was carried out (Steinberg et al. 2007). Collaborating with experts in Kings College London Dental School, researchers from hapTEL introduced the

Table 6.1 Comparison among existing dental simulators

System Specification	PerioSim	Simodont	hapTEL	Forsslund	VRDT
Force rendering dimension	3-DoF	3-DoF	3-DoF	3-DoF	3-DoF
Deformation simulation	NA	Tongue deformation	NA	NA	NA
Operation types	Periodontics	Endodontics	Endodontics	Endodontics	Endodontics
Oral environment	Upper/ lower jaw	One tooth	One tooth	One tooth	One tooth
Force feedback (FF)	Single-hand FF	Single-hand FF	Single-hand FF	Single-hand FF	Single-hand FF
Haptic–visual collocation	No collocation	Collocation	Collocation	Collocation	No collocation

evaluation method of two prototypes to identify design considerations including ergonomic factors and dental training needs. The authors summarized some areas meriting high priority, including finger rest, dental tools, software, and collocation among hand, eye, and tool (Tse et al. 2010).

Dental experts at ACTA carried out a study to investigate whether skills developed in virtual reality could be transferred to reality. They concluded that skills developed in virtual reality on the Simodont were transferrable to reality (Bakker 2009). Urbankova and Engebretson (2009) studied the relationship between performance on haptic exercise, PAT scores, and preclinical skill assessments. Their results provided evidence that haptic devices might play an important role in predicting performance in preclinical dental education. LeBlanc et al. compared virtual reality simulator-enhanced training with laboratory-only practice on the development of dental technical skills. The results indicated that students who trained with the virtual reality simulator between 6 and 10 h improved significantly more than the students in the control group (LeBlanc et al. 2004). Using probing and cavity preparation tasks, Konukseven et al. (2010) illustrated student's response on performance tests including usability, clarity, effectiveness, help/support provided, and satisfaction. Gal et al. (2011) showed that both experienced dental faculty members and advanced dental students found the simulator to have significant potential benefits in teaching manual skills in dentistry. Buchanan (2004) pointed out several challenges in virtual reality dental training, including the limitations of low numbers of units and hence low numbers of students, technical problems, and the desire to modify the educational environment.

The previous research results demonstrated that the virtual reality dental simulation systems could alleviate some problems in existing dental education, such as faculty shortages (Henzi et al. 2006). However, today's surgical simulators still cannot meet the stringent education requirements. Simulation fidelity of a simulator needs to be improved, and the value of simulation education needs to be validated by more randomized control studies. A good simulator should measure sufficient objective data about the trainee's performance and the tissue properties and thus provides quantitative foundations for evaluating training outcome.

6.2 Overview of iDental Simulation System

With over 10 years of collaboration between robotics researchers and dentists, we have developed iDental, a haptic–visual–audio feedback dental simulation system. Based on the configuration-based optimization approach for 6-DoF haptic rendering, the iDental simulator is capable of simulating multi-region contacts among surgical tools and tissues that possess complex geometric surfaces with fine geometric features.

In the following, the function requirements of dental training are first introduced, which provide a foundation for developing dental simulation systems. Next components and features of the simulator are introduced.

6.2.1 Function Requirements

To develop a practical dental simulator, we first need to define clear requirements including hardware and software and then map the requirements in each key step of medical training to system components of a dental simulator. The procedure for requirement analysis is as follows:

1. Select typical operations.
2. Identify key procedures.
3. Define function requirements for haptic and visual simulation.
4. Define performance specifications for each function item.

As shown in Fig. 6.1, three typical operations are considered in the iDental simulator: periodontics operation, endodontics operation, and prosthodontics operation.

In each operation, two to three key procedures are selected as training targets. For each key procedure, we first define function requirements for haptic and visual simulation and then define performance specifications for each function item.

The procedures of periodontal operation include three steps: (1) Use a periodontal probe to check the depth value of the pocket and thus to determine the degree of the periodontal inflammation, (2) use an explorer to detect the size, number, shape, and location of possible calculus, and (3) use scalers and curettes to remove calculus.

In the periodontal operation, haptic feeling is necessary to locate any invisible calculus inside a gingiva and to identify their shape and size. Haptic feeling is also needed to remove the calculus using appropriate angle of the tool and sufficient removal force.

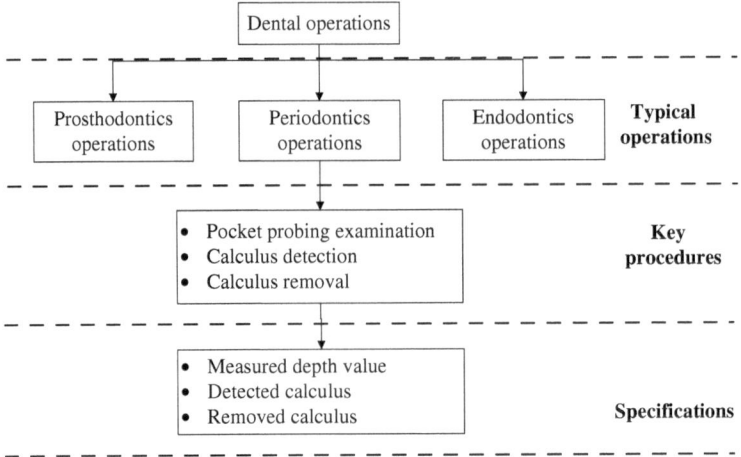

Fig. 6.1 Selected key procedures in *i*Dental simulator. Copyright © IEEE. All rights reserved. Reprinted, with permission, from Wang et al. (2012a, b, c)

For the task of measuring pocket depth, a dentist needs to maintain correct posture of the dental probe and avoid penetration with teeth and gingivae within a narrow space formed by oral cavity. For some teeth inside the oral cavity (such as tooth 46), it is a challenging task to obtain correct measurement without sufficient motor skill.

6.2.2 Components and Features of the Simulator

The components of the system include human trainee, two haptic devices, haptic–visual collocation platform, stereoscopic glasses, audio speakers, computer, simulation software, and evaluation software. Figure 6.2 illustrates the system components and a physical prototype of the system.

Two Phantom Omni devices are used as the haptic devices. The NVIDIA GeForce 3D Vision glasses are used. The purpose of the haptic–visual collocation platform is to realize spatial collocation between haptic display and visual display and thus to achieve a higher immersive feeling for fine manipulation (Wang et al. 2011). The specifications of the computer are as follows: Intel(R) Core(TM) 2 2.20 GHz, 2 GB memory, and X1550 Series Radeon graphic card. Note that no GPU is used.

The functions of the software include modeling of virtual tool and oral tissues, collision detection, collision response and force computation, and graphics display.

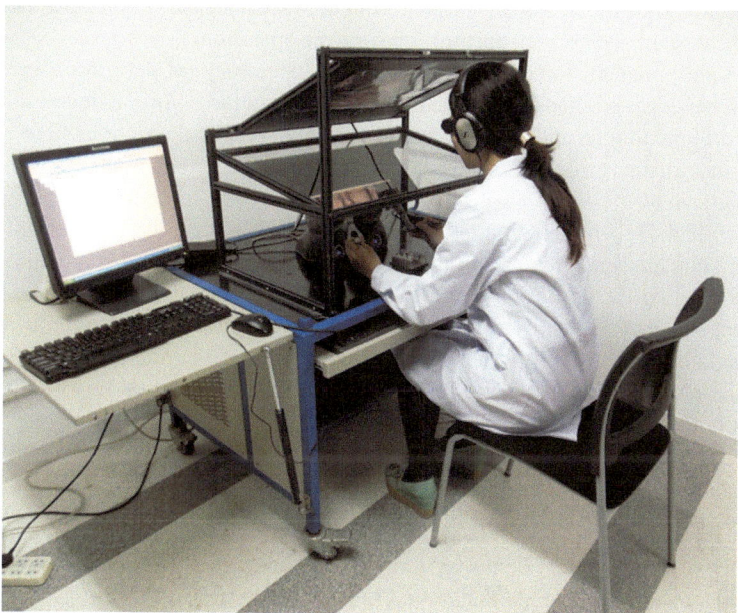

Fig. 6.2 The illustration of system components and a physical prototype of the iDental system

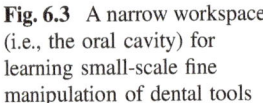

Fig. 6.3 A narrow workspace (i.e., the oral cavity) for learning small-scale fine manipulation of dental tools

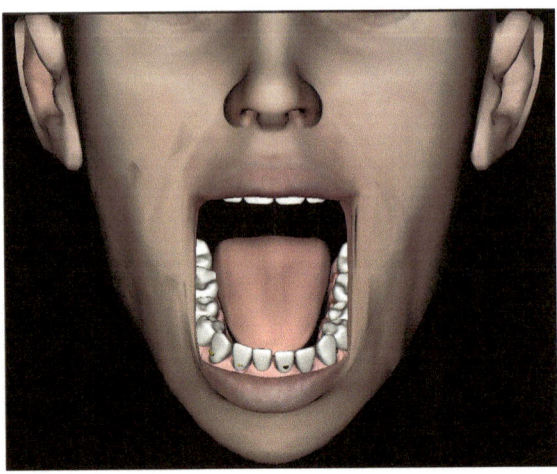

   The functions of the iDental system include virtual simulation and training for two typical periodontal operations, i.e., periodontal probing and calculus removal.

   Compared with existing dental simulation systems, the iDental system has the following novel features:

1. Provide a narrow workspace (i.e., the oral cavity) for learning small-scale fine manipulation of dental tools; as shown in Fig. 6.3, contact force can be felt through a haptic device when collision occurs between a tool and any tissues (including all the 28 teeth, gingiva, tongue, face/cheek). This feature is necessary to learn the correct posture for surgical operations.
2. Provide bimanual force feedback, which is necessary for learning coordinated task between two hands, such as using the left hand to control a dental mirror to deform the tongue and disclose a target tooth and using the right hand to control an instrument to detect or remove calculus of the target tooth (as shown in Fig. 6.4).
3. Provide 6-DoF haptic simulation; as shown in Fig. 6.5, the iDental prototype can simulate multi-region contacts between a dental tool and manipulated tissues (e.g., a tooth and its surrounding gingiva) (Wang et al. 2014). The tool is modeled as a rigid body instead of a point. Not only force but also torque can be simulated by a 6-DoF haptic device (e.g., Phantom Premium 3.0 6-DoF). For some tasks, learning to exert an expected torque on the dental tool is critical to the outcome of the surgical operation.
4. Provide haptic simulation of various physical properties, including rigid body contact, deformable object manipulation, and detection and removal of small-sized objects (as shown in Fig. 6.6).

   Other features of the iDental system include integration of haptic–visual collocation, stereoscopic display, and sound feedback. (as shown in Fig. 6.7).

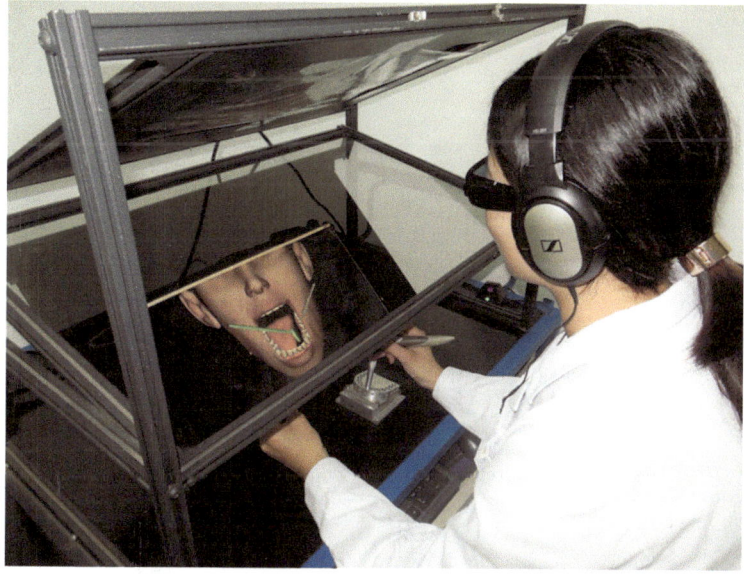

Fig. 6.4 Bimanual operation simulation for learning coordinated tasks between two hands

Fig. 6.5 6-DoF haptic
simulation for multi-region
contacts

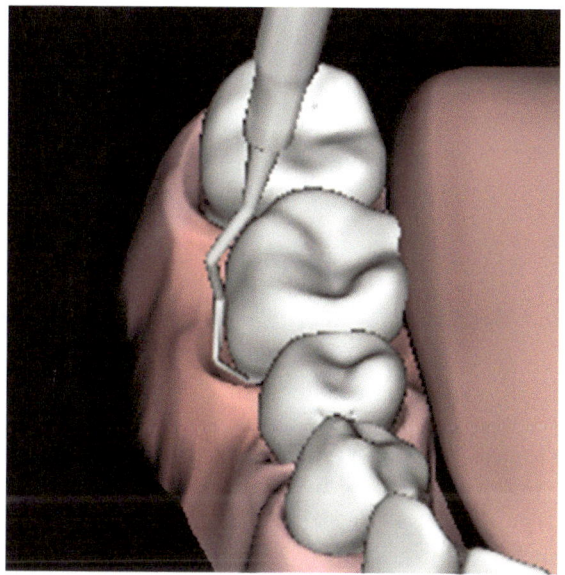

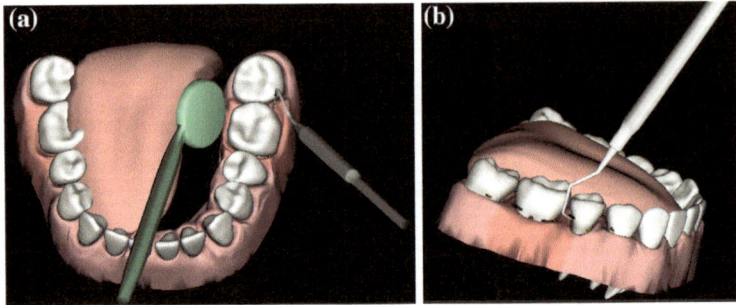

Fig. 6.6 Haptic simulation of various physical properties. **a** Deformable object manipulation and **b** detection and removal of small-sized objects

Fig. 6.7 Haptic–visual collocation to improve the feeling of immersion

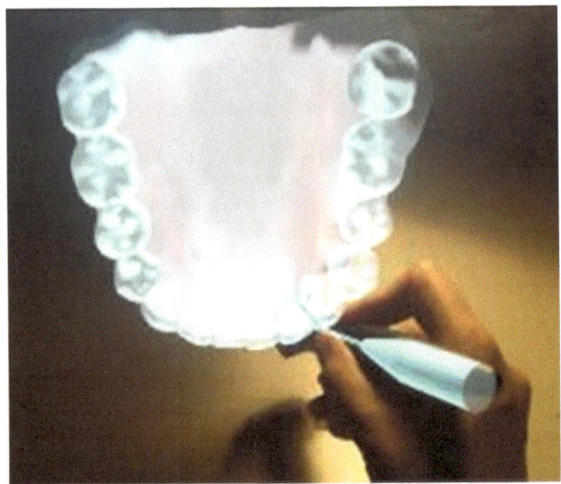

6.3 Modeling Various Tissues and Pathological Changes

Haptic simulation requires geometrical and physical modeling of the dental tools and the oral environment, including (1) modeling the geometric shapes and surgical behaviors of different dental instruments and (2) modeling pathological changes in terms of changes in geometric shapes and physical properties of gingiva, tooth, and calculus.

The teeth data used were derived from raw point cloud data recorded by a laser scanner and by CT. 3ds Max was used to refine the shape of a tooth and its neighboring gingiva and thus to produce the expected periodontal pocket with various depth and clearance width (Fig. 6.8). VRML was adopted as the import file format into the haptic scene.

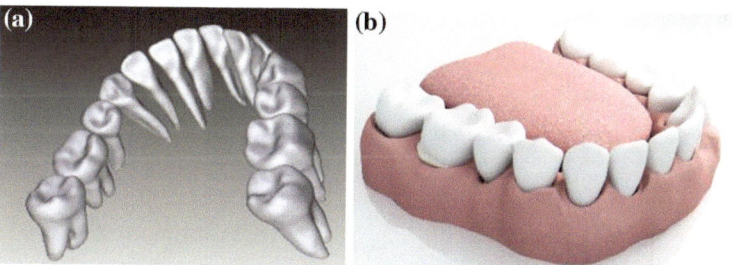

Fig. 6.8 **a** Original teeth data and **b** combined teeth and gingiva (including invisible sub-gingival calculus). Copyright © IEEE. All rights reserved. Reprinted, with permission, from Wang et al. (2012a, b, c)

The model for gingiva includes the following features: color, 3D shape, and material properties (such as stiffness and friction coefficient). These critical parameters can be tuned within a range, and a desired value for each parameter can be determined according to the subjective feeling of dental experts to reflect the degree of periodontal inflammation. The model for calculus includes the following features: shape, size, color, and location. Typical colors of a calculus include snuff and black colors.

As shown in Fig. 6.9, the shapes of various dental instruments were measured and modeled by using 3ds Max Studio (Autodesk Inc, USA). A sphere-tree toolkit was utilized to produce the haptic model for collision detection and collision response computation (see Chap. 2).

Blade directions of curettes were also defined based on the dental requirement. The working end of a curette was defined within 2 mm from the tip (i.e., in the segment A in Fig. 6.10) in order to avoid damaging soft tissues.

Fig. 6.9 3D models of various dental tools. From *left* to *right*: dental mirror, graduated Williams periodontal probe, explorer, and four types of Gracey curettes (13/14, 11/12, 7/8, 5/6)

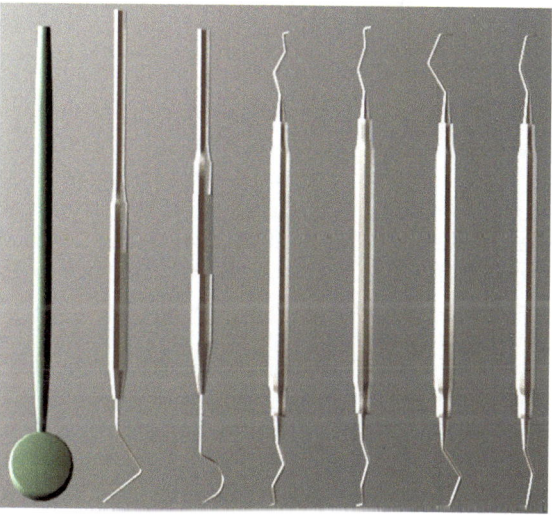

Fig. 6.10 Modeling surgical
behaviors of dental tools.
Copyright © IEEE. All rights
reserved. Reprinted, with
permission, from Wang et al.
(2012a, b, c)

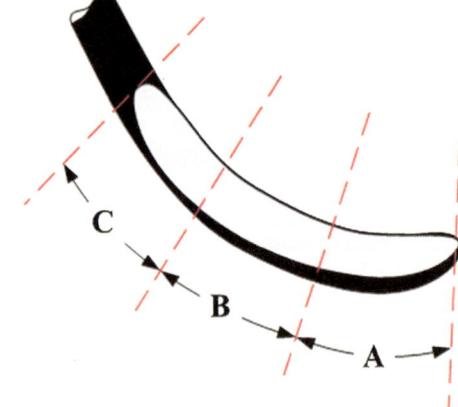

Fig. 6.11 Periodontal pocket
with different indicated depth
at six sites per tooth

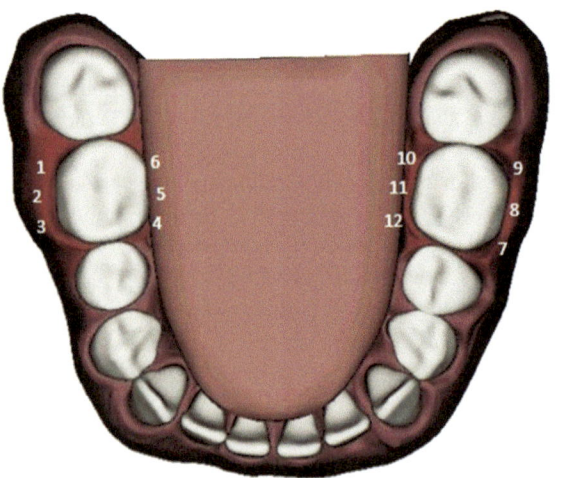

Different colors were used to represent inflamed gingiva as pathological changes. As shown in Fig. 6.11, the two teeth (i.e., tooth 36 and tooth 46) with dark red color represent inflamed gingiva. Furthermore, a periodontal pocket was defined for the tooth 36 and tooth 46, respectively. These two teeth were modeled with indicated deep pockets. The depth of the pocket could be randomly set from 2 to 8 mm. As shown in Fig. 6.10, six sites around the target tooth 46: mesio-buccal (3), central buccal (2), disto-buccal (1), mesio-lingual (4), central lingual (5), and disto-lingual (6). Similarly, six sites were selected for the target tooth 36.

Calculi with different shapes and sizes were modeled using the 3ds Max Studio and then transformed into the sphere-tree model introduced in Chap. 2. These calculi were placed randomly at different sites, while most of them were placed along the gingival margin. Figure 6.12 illustrates calculus at different locations, including both supra-gingival calculus and sub-gingival calculus.

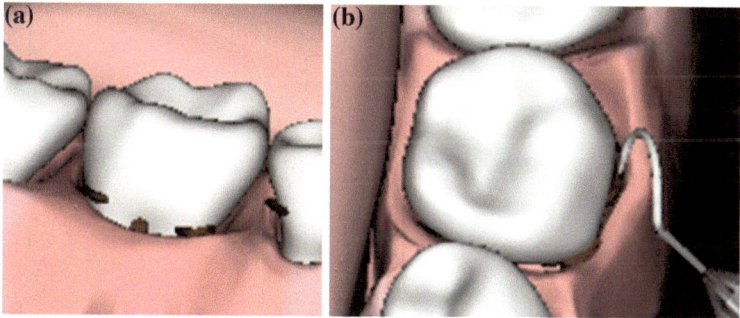

Fig. 6.12 Illustration of different calculi at different locations. **a** Supra-gingival calculus and **b** sub-gingival calculus

Parameters of material properties, such as stiffness and static and dynamic friction coefficients of different tissues, were tuned according to a dentist's subjective feeling.

Furthermore, the threshold for removal force was set randomly from 1 to 6 N. When a user exerts a force along a tooth's axis greater than the threshold, the calculus could be removed.

6.4 Bimanual Periodontal Operation Tasks

We developed two benchmark tasks: *periodontal probing* and *calculus removal*.

6.4.1 Periodontal Pocket Probing Examination

Periodontal pocket depth is an important indicator of periodontal inflammation, and its value denotes the degree of periodontal inflammation. Figure 6.13 shows the comparison between healthy periodontium and periodontium involved in inflammation.

In the *periodontal probing* task, the goal is to test whether a human trainee can manipulate two dental tools, a dental mirror and a probe, simultaneously. The human trainee is required to use his/her left hand to control the mirror to deform the tongue and right hand to measure the depth values of periodontal pockets around several hard-to-access teeth. There are two necessary motor skills involved in this task, bimanually coordinated manipulation to access a target pocket, and posture control of the two dental tools within a narrow space. Following data will be recorded as quantified metrics for evaluating the manipulation skill of trainees, including the times of unwanted collisions between the tool and oral tissues, the coordinated degree between the two tools, and the time to finish the required manipulation.

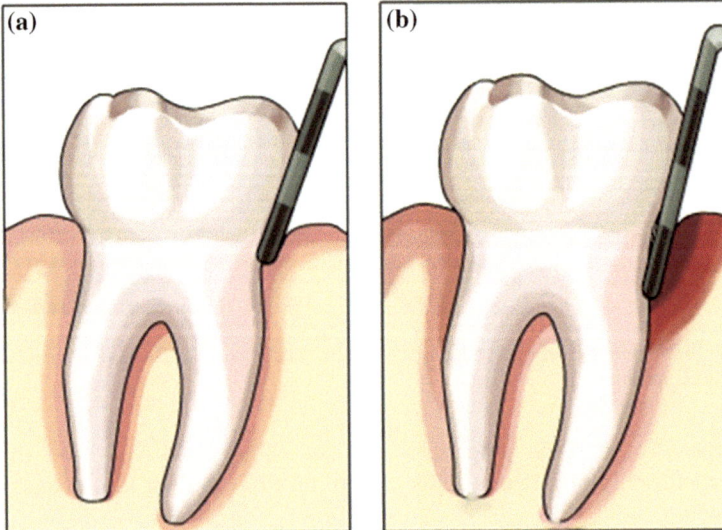

Fig. 6.13 a Healthy periodontium and **b** periodontium with inflammation

Fig. 6.14 Illustration of a periodontal pocket. Copyright © IEEE. All rights reserved. Reprinted, with permission, from Wang et al. (2012a, b, c)

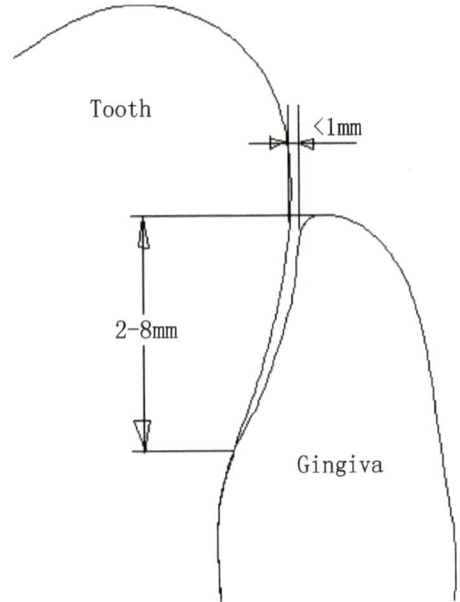

In the simulator, a periodontal pocket was modeled as a sulcus between a tooth and it surrounding gingiva (Fig. 6.14). The depth of the pocket could be randomly set in the range of 2–8 mm. A standard depth at each site was predefined in the 3D

geometrical model. This reference depth can be used to test the capability of the trainee, i.e., to test whether the trainee can tell the correct value of the depth.

A successful operation requires that (1) the probe is inserted into the bottom of a pocket, (2) the probe is kept parallel to the long axis of the tooth, (3) the probe's tip tightly contacts the surface of the root of the tooth and to push gingiva slightly to protect it from damage, and (4) the maximum resistance force at the bottom of the pocket is within 0.20–0.25 N.

During the simulation of periodontal pocket probing, an increasing force is simulated as the insertion depth of the tool increases. When the force is greater than the predefined force range, an audio feedback will be given as a warning and the event is recorded as an incorrect operation.

6.4.2 Calculus Detection and Removal

Calculus is mineralized or mineralizing bacterial plaque and debris, which is gradually deposited and formed on the surface of natural teeth and dental prostheses (the black volume in Fig. 6.15). Calculus is difficult to remove unless the calcified deposits are mechanically removed by a special instrument.

The surface of a normal tooth is smooth and continuous, but the surface of an unhealthy tooth is rough. The shape, location, and size of possible calculus are

Fig. 6.15 Illustration of calculus located below the marginal gingiva

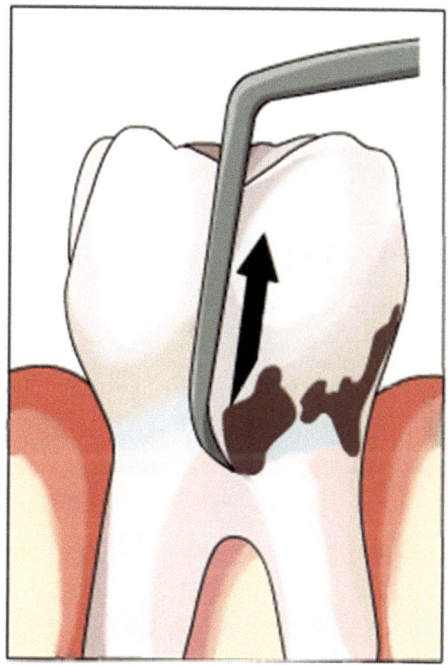

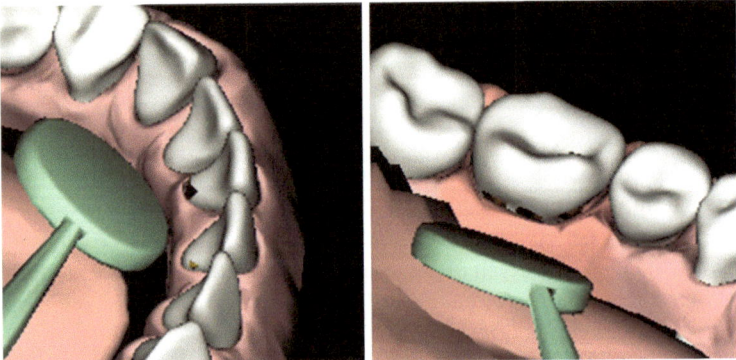

Fig. 6.16 The bimanual calculus removal task: A user uses his/her left hand to control the mirror to deform the tongue and uses right hand to access into some difficult accessible pockets and to realize detection and removal of invisible calculus

randomly chosen and placed in the haptic simulation environment. A human trainee is required to detect possible calculus hidden inside a pocket by tactile sensation only.

In a *calculus removal* task (shown in Fig. 6.16), a human trainee is required to use the left hand to control the mirror to deform the tongue and the right hand to insert the tool into some difficult to access pockets and to detect and remove invisible calculi, including supra-gingival calculus and sub-gingival calculus.

Successful calculus removal requires the dentist to find all the calculi deposited on the root of all teeth and remove them.

For calculus removal, there are strict restrictions on surgical parameters, including the type of surgical tool, blade direction of the tool, the orientation of the tool's axis, the threshold of removed force, and the valid (or safe) contact region between the tool and calculus. The angle between the tool's blade and the tooth surface is critical for removal. As shown in Fig. 6.17, the correct angle range is 70°–90° (80° is the best angle). If the angle is not correct, the tool will slip from the

Fig. 6.17 Correct angle range for calculus removal operation. Copyright © IEEE. All rights reserved. Reprinted, with permission, from Wang et al. (2012a, b, c)

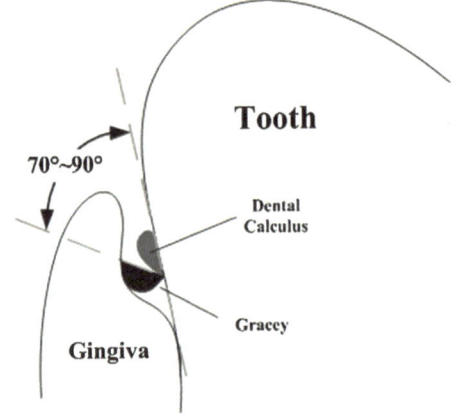

tooth surface, and the neighboring gingiva will be damaged. At the same time, the effective removal area on the tool is within the 1–2 mm close to the tip (i.e., in the segment A in Fig. 6.10), thus to avoid damaging soft tissues. Once the calculus is removed, the dentist feels a sudden drop of force magnitude.

There are delicate tools, such as scalers and curettes, for removing supra-gingival calculus and sub-gingival calculus. For teeth at different areas, the removal tools can be different—different curettes are used for treating molars and fore teeth.

In the simulation of calculus removal, all the calculi are assumed to be block-shaped. Angle and removal force threshold are two simulation parameters.

6.5   Hybrid Evaluation Approach

A combined approach was used, which consists of subjective and objective evaluations, to evaluate the performance of the iDental system (Wang et al. 2012a, b, c).

6.5.1   Subjective Evaluation Method

Questionnaires for dentists of various skill levels were used to gather subjective and qualitative evaluation data. Each subject was instructed to use a structured global assessment scale to provide evaluation score on the haptic fidelity of the system (Lee et al. 2009). Realism scores on the haptic feeling span seven levels, i.e., scores from 1 to 7, where 7 refers to high fidelity, 4 refers to medium fidelity, and 1 refers to low fidelity.

Figure 6.18 shows the qualitative evaluation architecture (Fig. 6.18), which was designed based on the analysis of function components of the iDental system.

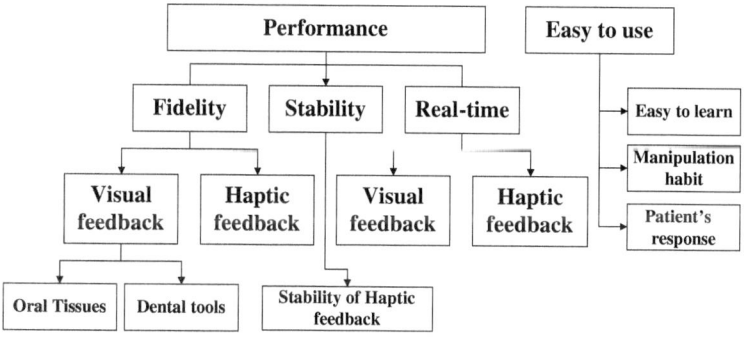

Fig. 6.18   A qualitative evaluation architecture for dental simulator. Copyright © IEEE. All rights reserved. Reprinted, with permission, from Wang et al. (2012a)

The details of several key components are as follows:

- Fidelity of visual feedback (mainly evaluation on the shape and color of tissues/tools)

  - shape and color of teeth/gingiva/calculus
  - shape and color of dental instruments
  - shape and color of oral environment, including cheek and tongue.

- Fidelity of haptic feedback when the tool contacts tissues

  - Operation during pocket probing check, including magnitude and direction of force, allowable moving range and allowable orientation range of the dental probe, and stiffness and friction of the tissues (teeth and gingiva)
  - Operation during calculus removal, including magnitude and direction of force, allowable moving range and allowable orientation range of the dental probe, and stiffness and friction of the tissues (teeth and gingiva)

- Convenience degree for use and learn

  - Ease of use: whether the user interface is easy for dentist to learn and get familiar with the system;
  - Comparable clinical experience: how the experience is different from clinical cases, including firm finger rest, vision field, bimanual simultaneous operation habit, and force magnitude.
  - Response of virtual patient: how the patient responds to illegal operations.

6.5.2 Objective Evaluation Method

Haptic feedback is an important component in a dental simulator. However, there are many open questions, including which performance metrics of haptic feedback are useful for dentist training and which are not necessary, as well as how to quantify the effect of those useful performance metrics.

An objective and quantitative evaluation method was introduced to address these questions, which relies on identifying key parameters from the target task and defining quantified metrics for those parameters. Based on those parameters, the concept of construct validity was used to compare performance among different subjects of the same simulator (Ullrich and Kuhlen 2012), i.e., to detect whether significant differences exist between experts and novices.

To develop a practical surgical simulator, the metrics should be defined in a task analysis preceding the simulator development and not as an afterthought.

For each trainee, quantifiable data are collected for each benchmark task. For the *periodontal probing* task, the measured and collected data are (1) measured value of the depth of the target pocket, (2) number of unwanted collisions between the tool and oral tissues, and (3) time to finish the required manipulation. For the *calculus removal* task, the measured and collected data are (1) number of successfully removed calculi, (2) number of unwanted collisions between the probe's tip and the tongue and cheeks, (3) the time to finish the required manipulation, and (4) other data, including deformation value of the tongue, force/torque data between the probe and the pocket, and force/torque data between the mirror and the tongue.

For each tooth with six sites of pocket around the tooth (Fig. 6.10), the depth value at each site is predefined as a constant value, which was denoted as a standard value. The difference between the recorded depth value and the standard depth value reflects the skill level of the subject. Furthermore, the maximum force when the probe contacts the bottom of the pocket was also used as an indicator of the skill level. This force should be between 0.20 and 0.25 N. Finally, the angulation between the probe and the tooth axis also reflects the skill level.

6.6 User Study and Evaluation Results

Three user groups participated in evaluating the performance of the iDental simulation system, including an expert-level group (8 dentists with over 10 years of clinical practices, 2 males and 6 females), a medium-level group (9 residence dentists with 3–10 years of clinical practices, 3 males and 6 females), and a novice-level group (14 graduate students with less than 3 years of clinical practices, 7 males and 7 females).

6.6.1 Subjective Evaluation Data

Figure 6.19 shows the average evaluation data of the three groups in terms of 1–7 of a global assessment scale for subjective feeling of realism of force feedback. Most of the values except for the gingiva contact are greater than four, which means that the force fidelity was considered above the medium fidelity. The gingiva contact and the hybrid contact with tooth and the gingiva are the two cases having lower scores. Note that the case of multi-region contacts has a high score, and most subjects think that the simulation of multi-region contacts provides more immersive feeling than single-region contacts.

Subjective evaluation data on the fidelity of graphics display are shown in Fig. 6.20. Again, the average values for all cases are greater than four.

Further evaluation is performed to validate the added benefit of new features including the narrow space simulation and deformation simulation. Subjective evaluation data on these features are shown in Fig. 6.21. The average values of most metrics except for the orientation accuracy of the virtual tool are greater than four. From the results, most subjects think that narrow space simulation provided an immersive feeling and is effective to simulate the challenge in practicing fine

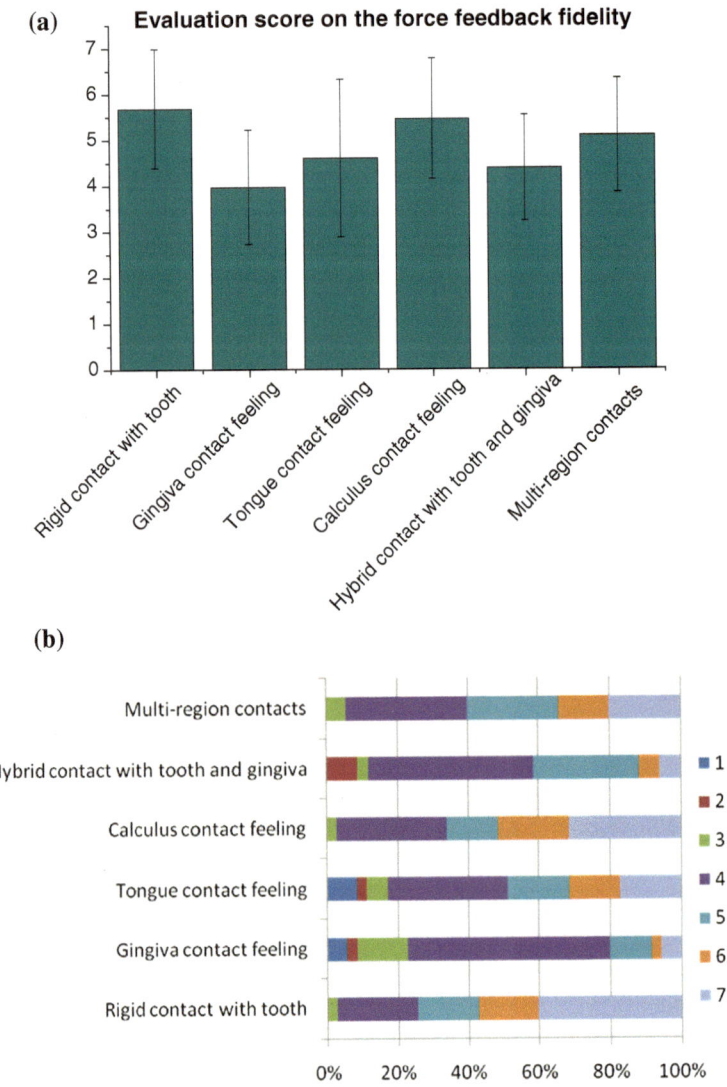

Fig. 6.19 Evaluation score on the force feedback fidelity during contacts of the tool with various tissues. **a** Mean and standard deviation of the score and **b** distribution of participants' number at each score level

instrument manipulation. Most subjects put high priority on the necessity of simulating a deforming cheek, because this is necessary to probe a tooth that locates in the side and occluded by the cheek. The orientation accuracy of the virtual tool is not satisfied, which is caused by the low sensing accuracy of the haptic device.

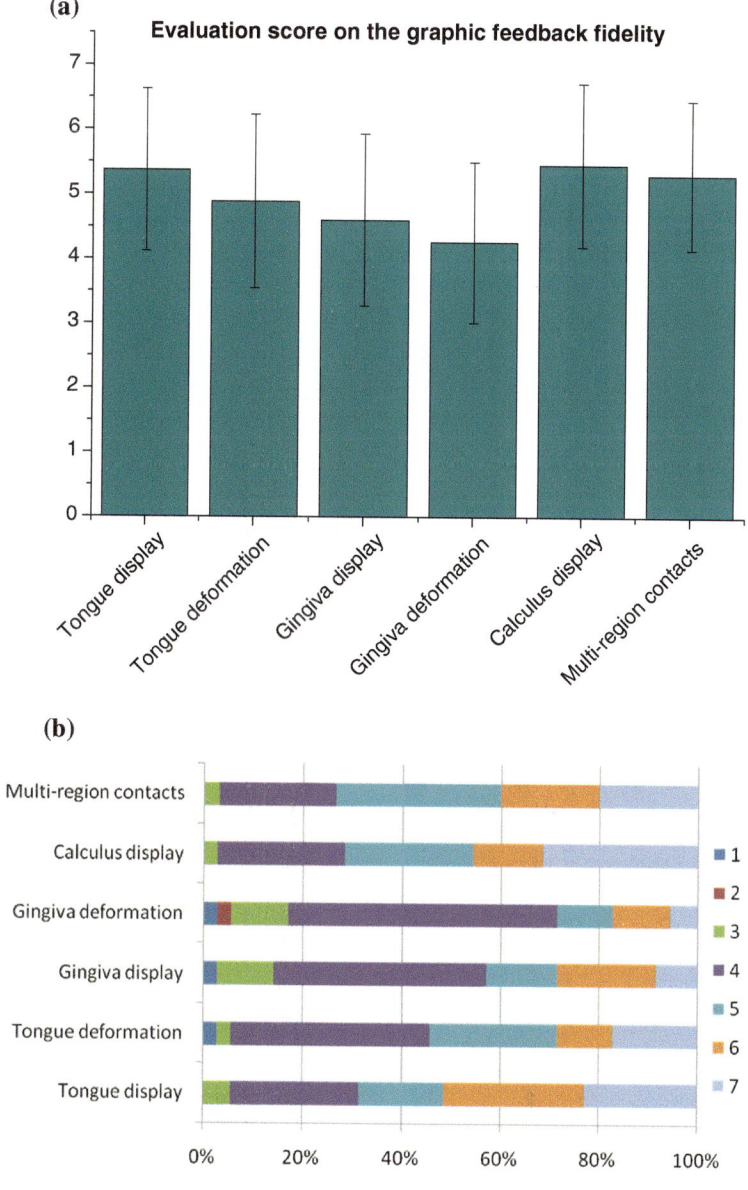

Fig. 6.20 Evaluation score on the fidelity of graphics display during contacts of the tool with various tissues. **a** Mean and standard deviation of the score and **b** distribution of participants' number at each score level

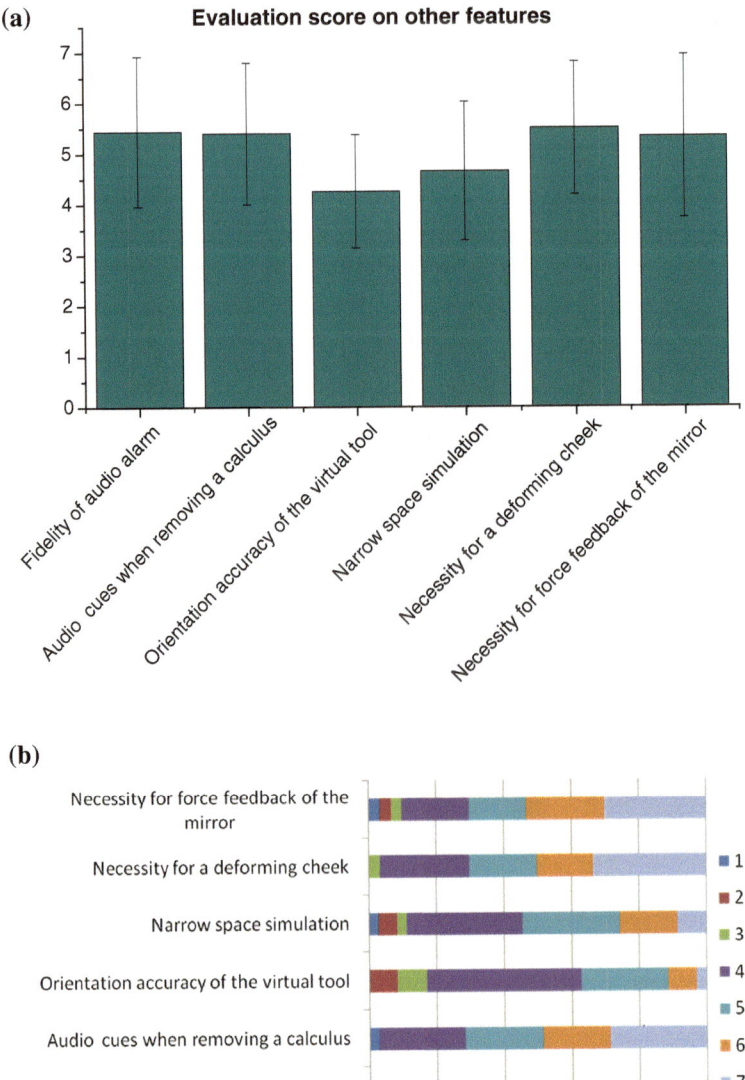

Fig. 6.21 Evaluation score on other features. **a** Mean and standard deviation of the score and **b** distribution of participants' number at each score level

6.6.2 *Quantified Data*

For the *periodontal probing* task, Table 6.2 illustrates the measured probing depth at the 12 indicated sites (Fig. 6.11). For each group, the average depth and the standard deviation values are computed. Assuming a dental expert will maintain a correct posture between the two tools, and the depth value on the probe can thus be read correctly through the mirror, while a novice will read an incorrect value because the relative angle between the mirror and the probe is not correct. Based on our assumption, the expert group is expected to produce small standard deviation, which reflects not only the fidelity of the simulation system, but also their high skill level.

In each row of Table 6.2, the bold number represents the group that has the smallest standard deviation. For site 5, 9, 10, 11, and 12, the expert group manifested the smallest standard deviation between the three groups, which implied that their measurement results were consistent.

The above results could be confirmed by observing simulation fidelity at those sites. As shown in Figs. 6.22 and 6.23, the simulation fidelity of pocket probing appears high at sites 9 and 12, in that the deformation of the gingiva in simulation is similar to that in the real operation scenario using a plastic model.

For the other seven probe sites, the expert group produced a greater standard deviation, and the fidelity of pocket probing simulation at these sites is low. As shown in Fig. 6.24a, at site 2 of the tooth 46, the simulated depth of the pocket is too shallow, and thus, it is difficult to insert the probe into the pocket. In the real operation scenario shown in Fig. 6.24b, the probe can easily be inserted into the pocket of the plastic model.

As shown in Fig. 6.25a, at site 4 of the tooth 46, the lateral deformation of the periodontal pocket is too large compared to the real scenario, which produced

Table 6.2 Measured probing depth of 12 indicated sites

Site	Expert level		Medium level		Novice level	
	Mean	Std.	Mean	Std.	Mean	Std.
1	2.8	1.6	3.1	**1.3**	3.2	1.5
2	2.4	1.3	2.6	1.3	2.9	**1.2**
3	2.0	1.1	2.4	1.1	2.4	**0.7**
4	2.9	1.2	2.9	1 4	2.9	**1.1**
5	2.4	**0.9**	3.1	1.5	2.5	1.0
6	3.0	1.4	3.2	**0.8**	2.6	1.1
7	4.6	0.7	4.8	**0.7**	3.7	1.3
8	2.9	1.2	3.6	1.4	2.6	**0.9**
9	4.5	**0.9**	5.1	2.8	4.7	1.9
10	3.6	**1.3**	4.8	1.8	3.9	1.4
11	4.4	**1.2**	4.2	1.9	4.2	1.6
12	3.5	**1.1**	4.0	1.9	3.6	1.6

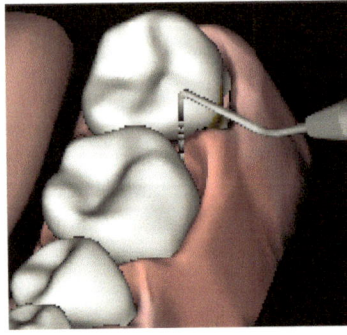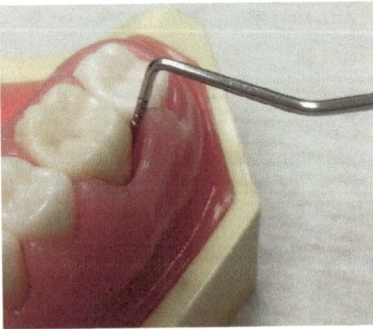

Fig. 6.22 High fidelity of pocket probing simulation at site 9 of tooth 36. *Left* simulated system, *right* real system

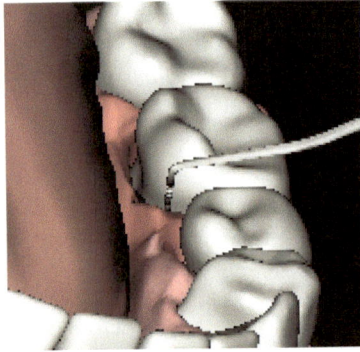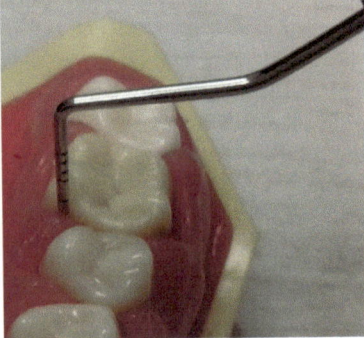

Fig. 6.23 High fidelity of pocket probing simulation at site 12 of tooth 36. *Left* simulated system, *right* real system

ambiguity for reading the depth of the pocket and was difficult for the human subject to judge the depth of the periodontal pocket.

Furthermore, in the simulation system, the bottom is formulated by sphere trees of the gingiva, which produced approximation errors of the bottom shape.

The average number of incorrect contacts between the probe and the tongue are shown in Fig. 6.26 for the three user groups. The average numbers of the expert and the novice groups are smaller than those of the medium-level group. The reason is that users in the novice group were very cautious during their operation process, while users in the expert group can control the probe skillfully. For the group in the middle, the users did not pay much attention to the operation, and therefore, the average number of incorrect contacts is large. The total operation time costs of the three groups are shown in Fig. 6.27. There is no significant difference between the three groups.

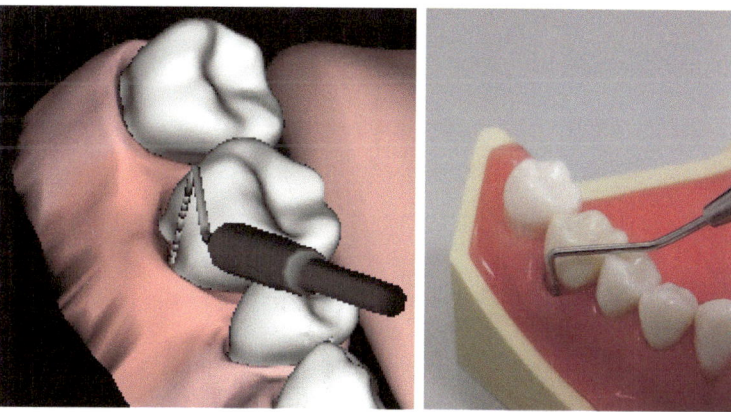

Fig. 6.24 Low fidelity of pocket probing simulation at site 2 of tooth 46. *Left* simulated system, *right* real system

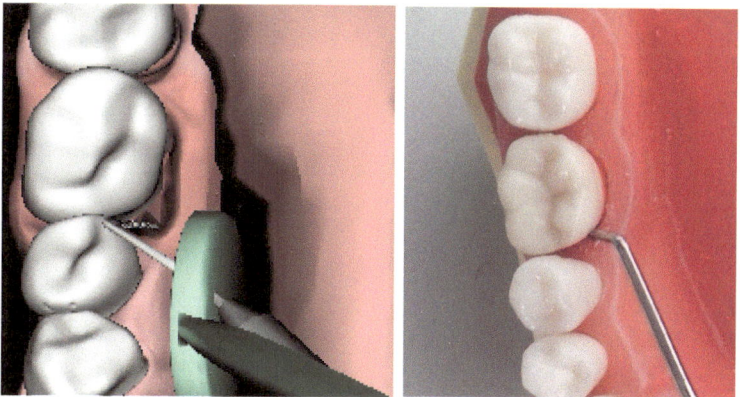

Fig. 6.25 Low fidelity of pocket probing simulation at site 4 of tooth 46. *Left* simulated system, *right* real system

For the *calculus removal* task, the number of successfully removed calculi and the time cost of removing a single calculus is shown in Fig. 6.28. We can see that almost all subjects could detect the calculi and remove them, while the expert group used less time than the other two groups. The possible reason is that the experts are more familiar with the anatomy of the teeth and thus can detect the locations of the calculus more quickly.

The number of incorrect contacts between the probe and the tongue is shown in Fig. 6.29 for the three groups. The average number of the novice group is smaller than that of the expert and medium groups. This result seems to contradict

Fig. 6.26 Average number of incorrect contacts during the pocket measurement task

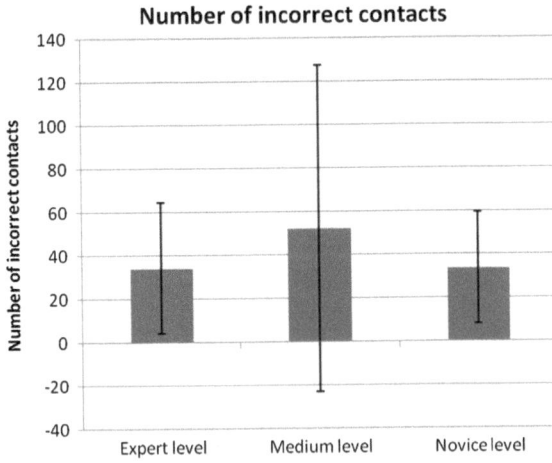

Fig. 6.27 Total time costs of the periodontal probing task

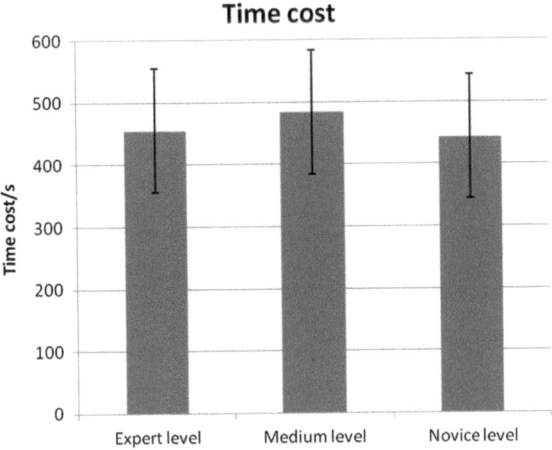

Fig. 6.28 Average number of successfully removed calculi

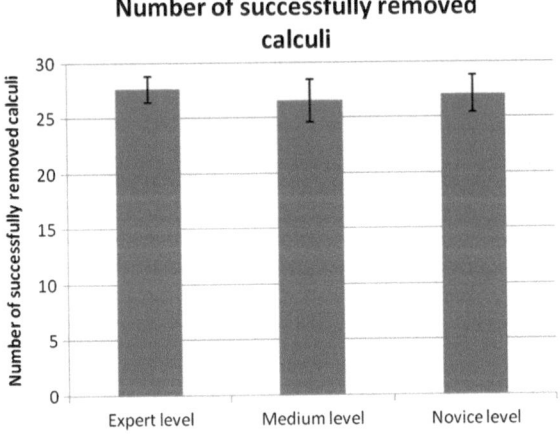

Fig. 6.29 Average number of incorrect contacts during the calculus removal task

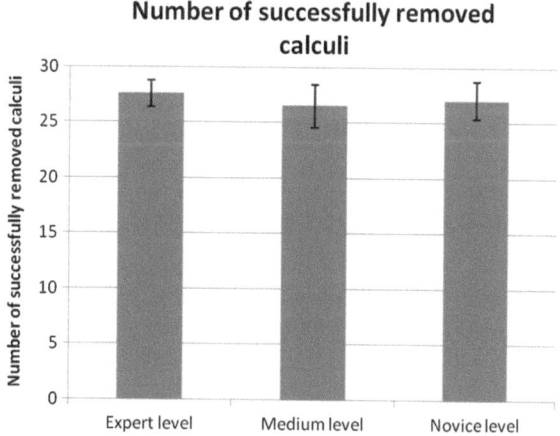

the common sense. Our conjecture is that a dental expert moved the two tools simultaneously to their target location to be efficient, while a novice moved one tool at a time and therefore produced fewer incorrect results, i.e., the time cost data shown in Fig. 6.30 support the conjecture. The total operation time cost of the novice group is greater than that of the expert and medium groups. The students spent more time to detect the calculus, and they moved cautiously in the virtual mouth while trying their best to avoid incorrect contacts between the probe and the tongue.

By comparing the number of incorrect contacts and time costs, it seems that the simulation system fails to simulate the real system with the necessary fidelity to make experts feel comfortable. In another aspect, reluctance to practice on the surgical simulator affects the performance of simulated operations. Those subjects who spent more time to get familiar with the system could better manipulate the haptic device to access the target tooth and avoid unnecessary contacts with the tongue. On the other hand, users in the expert group were reluctant to spend enough time to learn the new system. Therefore, they lacked spatial perception in the simulated environment. This is the main reason why the scores of the novice group are better than those of the

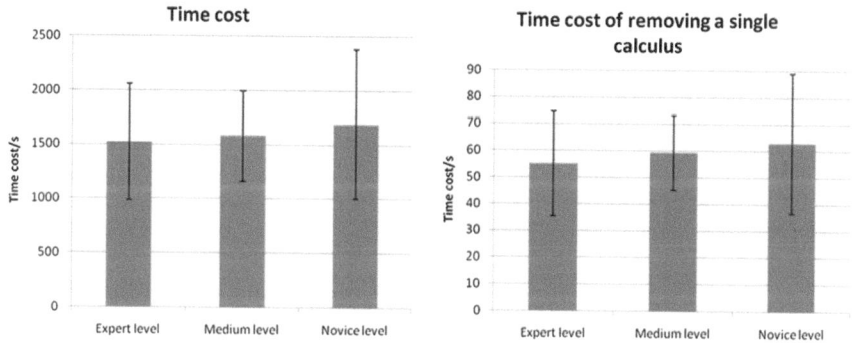

Fig. 6.30 Time costs of removing all calculi and mean time for a single calculus

medium group, because most graduate students spent sufficient time (i.e., the required learning time period of 40 min) to adapt to the haptic device.

Furthermore, it is more difficult for experts to adapt to the new manipulation habit of the haptic device than novices.

6.7 Lessons from User Study

Based on the evaluation results from the user study, important lessons can be identified to improve the hardware and software of the simulation system.

6.7.1 Needed Improvement in the Hardware System

One lesson is that dental experts did not score better than novice users in using the simulation system, which may indicate that the current simulator cannot capture accurately certain skills in actual clinical operations.

In order to provide a high-fidelity simulation for calculus removal, it is necessary to improve the performance of the haptic device. As shown in Fig. 6.31, the diameter of the Phantom Omni device is much greater than the diameter of a periodontal probe. This difference makes it difficult for a dentist to transfer the manipulation skills obtained from operating the real probe to the Phantom device used by the simulator, since the feeling of grasping a thick shaft is different from gripping a thin shaft. Moreover, unlike in actual operation, where a dentist usually uses the adjacent tooth of the target tooth as the operation fulcrum, to operate the haptic device, the operator has to use a tooth far from the target tooth as the supporting fulcrum. The much increased distance between the operation fulcrum and the target tooth could make the fine manipulation infeasible.

Therefore, there is a need to develop new haptic devices to meet the specific requirements of dental operations. First, the generic stylus of the haptic device

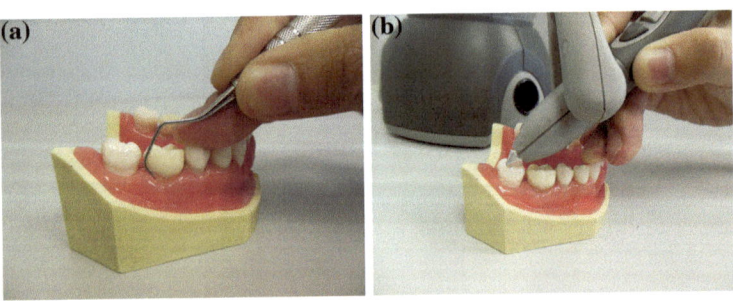

Fig. 6.31 Comparison of tool-operating styles between clinical scenario and simulation. **a** The clinical scenario and **b** the simulation system

Fig. 6.32 Two phantom Omni devices cannot operate closer to each other to simulate the needed bimanual coordination

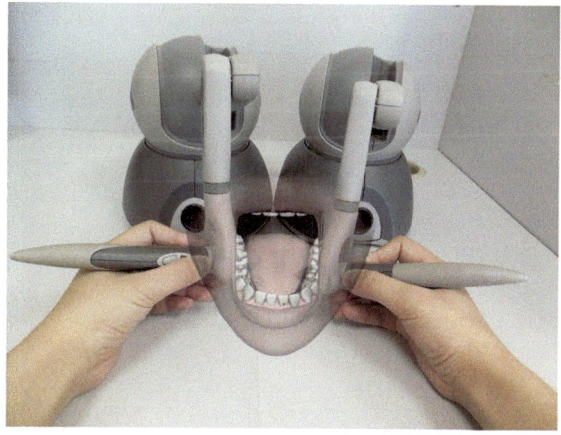

should be replaced a real dental instrument. The haptic device should be able to allow interchangeable styluses to adapt to various tools for dental operations.

Furthermore, it is necessary to increase the maximum achievable stiffness of the haptic device. In fine manipulation, such as removing a calculus, the movement range of the dental tool's tip is about 1–2 mm, and the removal force is about 10–20 N. Therefore, a haptic device should provide a stiffness of about 10–20 N/mm. To achieve stiff feeling of probing against enamel, the stiffness should be over 100 N/mm. However, the maximum achievable stiffness of the Phantom Omni device is only about 1 N/mm throughout the workspace of the device. Because it is difficult to achieve a greater stiffness for a device with impedance display, one possible solution is to use devices with admittance display, such as the Haptic Master developed by MOOG Inc.

Another requirement is that the footprint or the size of the haptic device needs to be small, as the tips of the two devices need to be placed very close, i.e., within the real size of a patient's mouth. Otherwise, the fidelity for training bimanual operations will be reduced. As shown in Fig. 6.32, it is difficult to put the two Phantom Omni devices to be close enough to simulate the bimanual coordination within a mouth cavity, because the mechanical structures of the two styluses and/or the bases will interfere with each other.

6.7.2 Needed Improvement in the Software System

Based on the user's feedback, the following aspects of the simulation software need to be improved:

1. Tool–tool collision simulation: The current simulation system does not conduct collision checking between two tools in a bimanual operation. As the result, the mirror and the probe in Fig. 6.33 could penetrate into each other.

Fig. 6.33 The mirror and
probe may penetrate into each
other

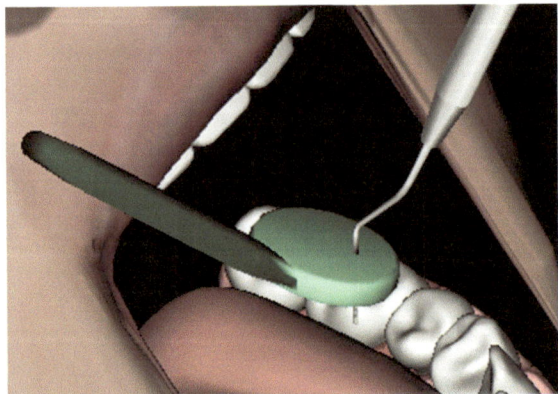

Fig. 6.34 Deformation of the
cheek caused by pulling with
a dental mirror

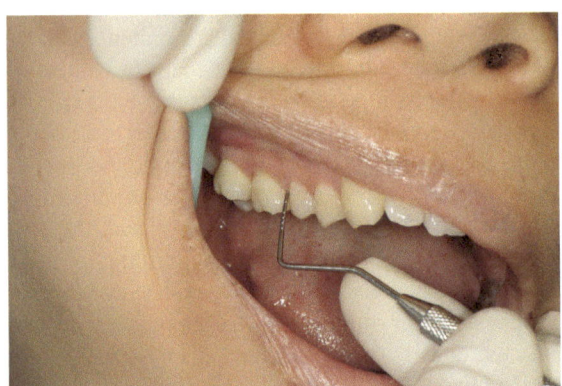

2. Cheek deformation: In order to inspect and remove calculi far inside a mouth, it
   is necessary to simulate the deformation of the cheek to allow the cheek being
   pulled a dental mirror, as shown in Fig. 6.34. This requires extension of the
   current simulation system to enable simulating physically realistic large-scale
   deformation of the cheek.
3. Simulation of handling rare cases of periodontal diseases: As the physical
   characteristics of a gingiva will be different under various pathological states,
   more versatile behaviors of a gingiva need to be simulated. Furthermore, more
   types of calculi need to be simulated such as particle-shaped calculi. Simulation
   of these pathological changes is necessary to demonstrate the advantages of
   virtual training methods over traditional training based on a phantom head.

6.8 Summary

In this chapter, the simulator iDental is introduced for periodontics operations. The prototype is used to understand the role of haptics in dental simulation and identify useful lessons. Compared to existing dental simulators, the unique features of the iDental system include 6-DoF haptic feedback, deformable objects' simulation, bimanual coordination training, and simulation of fine manipulations in a narrow oral cavity.

Evaluation results of iDental are also presented. Based on the results of user evaluation, needed improvements for hardware and software are identified, which point to future research and development topics for producing better dental simulators based on haptics.

References

Aggarwal R, Darzi A (2009) From scalpel to simulator: a surgical journey. Surgery 145:1–4
Bakker D (2009) Acceptance of the Simodont as a virtual training system. In: Proceedings of 35th annual meeting of the association for dental education in Europe (ADEE 2009), Helsinki, Finland, 26–29 Aug 2009
Basdogan C, Sedef M, Harders M, Wesarg S (2007) VR-based simulators for training in minimally invasive surgery. IEEE Comput Graph Appl 27:54–66
Buchanan JA (2004) Experience with virtual reality-based technology in teaching restorative dental procedures. J Dent Educ 68(12):1258–1265
Coles TR, Meglan D, John NW (2010) The Role of haptics in medical training simulators: a survey of the state-of-the-art. IEEE Trans Haptics, pp. 51–66
Crossan A (2003) The design and evaluation of a haptic veterinary palpation training simulator. PhD dissertation, Department of Computing Science and Faculty of Veterinary Medicine, University of Glasgow, Dec 2003
Forsslund J, Sallnas EL, Palmerius KJ (2009) A user-centered designed FOSS implementation of bone surgery simulations. In: World haptics conference, pp 391–392
Gal GB, Weiss EI, Gafni N, Ziv A (2011) Preliminary assessment of faculty and student perception of a haptic virtual reality simulator for training dental manual dexterity. J Dent Educ 75 (4):496–504
Henzi D, Davis E, Jasinevicius R, Hendricson W (2006) North American dental students' perspectives about their clinical education. J Dent Educ 70(4):361–377
Jasinevicius TR, Landers M, Nelson S, Urbankov A (2004) An evaluation of two dental simulation systems: virtual reality versus contemporary non-computer-assisted. J Dent Educ 68 (11):1151–1162
Konukseven EI, Onder ME, Mumcuoglu E, Kisnisci RS (2010) Development of a visio-haptic integrated dental training simulation system. J Dent Educ 74(8):880–891
Kothari SN, Kaplan BJ, DeMaria EJ, Broderick TJ, Merrell RC (2002) Training in laparoscopic suturing skills using a new computer-based virtual reality simulator (MIST-VR) provides results comparable to those with an established pelvic trainer system. J Laparoendosc Adv Surg Tech 12:167–173
LeBlanc VR, Urbankova A, Hadavi F, Lichtenthal RM (2004) A preliminary study in using virtual reality to train dental students. J Dent Educ 68(3):378–383

Lee Jason T et al (2009) The utility of endovascular simulation to improve technical performance and stimulate continued interest of preclinical medical students in vascular surgery. J Surg Educ 66(6):367–373

Luciano C, Banerjee P, DeFanti T (2009) Haptics-based virtual reality periodontal training simulator. Virtual Reality 13(2):69–85

Quinn F, Keogh P, McDonald A, Hussey D (2007) A pilot study comparing the effectiveness of conventional training and virtual reality simulation in the skills acquisition of junior dental students. Eur J Dent Educ 7(4):13–19

Ranta JR, Aviles WA (1999) The virtual reality dental training system—simulating dental procedures for the purpose of training dental students using haptics. In: Proceedings of the 4th PHANTOM users group workshop, Nov 1999

Steinberg D, Bashook PG, Drummond J, Ashrafi S, Zefran M (2007) Assessment of faculty perception of content validity of PerioSim(C), a haptic-3D virtual reality dental training simulator. Dent Educ 71(12):1574–1582

Thomas G, Johnson L, Dow S, Stanford C (2000) The design and testing of a force feedback dental simulator. Comput Methods Programs Biomed 64(1):53–64

Tse B, Harwin W, Barrow A, Quinn B, San Diego J Cox M (2010) Design and development of a haptic dental training system—hapTEL. In: EuroHaptics 2010 conference, VU University, Amsterdam, The Netherlands. Lecture notes in computer science, vol 6192/2010, pp 101–108

Ullrich S, Kuhlen T (2012) Haptic palpation for medical simulation in virtual environments. IEEE Trans Vis Comput Graph 18(4):617–625

Urbankova A, Engebretson S (2009) The use of haptic technology to predict preclinical dentistry performance, and perceptual ability: a novel method for identifying non-cognitive skill development potential. In: Proceedings of 35th annual meeting of the association for dental education in Europe (ADEE 2009), Helsinki, 26–29 Aug 2009

Wang D, Zhang Y, Zhou W, Zhao H, Chen Z (2011) Collocation accuracy of visuo-haptic system: metrics and calibration. IEEE Trans Haptics 4(4):321–326

Wang D, Zhang Y, Hou J, Wang Y, Lü P, Chen Y, Zhao H (2012a) iDental: a haptic-based dental simulator and its preliminary evaluation. IEEE Trans Haptics 5(4):332–343

Wang D, Liu S, Zhang X, Zhang Y, Xiao J (2012b) Six-degree-of-freedom haptic simulation of organ deformation in dental operations. In: IEEE international conference on robotics and automation (ICRA 2012), St. Paul, pp 1050–1056, 14–18 May 2012

Wang D, Liu S, Xiao J, Hou J, Zhang Y (2012c) Six degree-of-freedom haptic simulation of pathological changes in periodontal operations. In: IEEE/RSJ international conference on intelligent robots and systems (IROS2012), Vilamoura, 7–12 Oct 2012

Wang D, Shi Y, Liu S, Zhang Y, Xiao J (2014) Haptic simulation of organ deformation and hybrid contacts in dental operations. IEEE Trans Haptics 7(1):48–60

Chapter 7
Conclusions and Future Work

6-DoF haptic rendering for fine manipulation in narrow space is a challenging topic. This is due to frequent constraint changes caused by small tool movement and the need to preserve the corresponding force feeling when the tool interacts with fine features of an object. In this book, a novel constraint-based approach, i.e., the configuration-based optimization approach, has been introduced to tackle the challenge. In this chapter, we summarize the book by briefly discussing open challenges and possible research topics in this exciting field.

Each virtual object as well as the virtual tool is represented by a hierarchy of spheres, i.e., a sphere tree, which allows faster detection of multiple contacts/collisions among objects than a polygonal mesh and facilitates contact constraint formulation.

Given the pose of the haptic tool in a virtual environment, the configuration of the corresponding graphic tool, which does not penetrate into virtual objects, is computed by solving a configuration-based optimization problem. The constraints in the 6D configuration space of the graphic tool are obtained and updated through online mapping of the non-penetration constraints between the spheres of the graphic tool and those of the objects in the 3D virtual environment, based on the result of collision detection. This problem is further modeled as a quadratic programming optimization and solved by the classic active-set methods.

The approach has been extended to simulating objects with fine geometric features, including sharp features and small-sized geometric features, and multi-region frictional contacts. Sharp features are modeled by using a multi-resolution sphere tree which is formulated by splitting spheres in the neighborhood of a sharp edge. Furthermore, the approach has been extended to simulating deformation and hybrid contacts by defining an extended sphere-tree model with springs and dampers for modeling deformable objects.

The configuration-based constrained optimization approach has been implemented and interfaced with a 6-DoF Phantom Premium 3.0. Its performance is demonstrated in several benchmarks involving complex, multi-region contacts. The experimental results show both high efficiency and high stability of haptic rendering

© Springer-Verlag Berlin Heidelberg 2014
D. Wang et al., *Haptic Rendering for Simulation of Fine Manipulation*,
DOI 10.1007/978-3-662-44949-3_7

of complex scenarios. Non-penetration between the graphic tool and objects is maintained under frequent contact switches. The update rate of the simulation loop is maintained at about 1 kHz.

Evaluation of haptic rendering algorithms is an important topic toward achieving realistic haptic simulation. A measurement-based method is introduced to evaluate the accuracy of different 6-DoF haptic rendering algorithms for interactions involving multi-region contacts.

Based on the configuration-based optimization approach, a haptic-enabled dental simulator, iDental, is developed as an application to test the efficacy of the approach in simulation of fine manipulation. Preliminary results of user evaluation studies consisting of subjective and objective evaluations validate the effectiveness of the approach.

Besides the work in this book, there are still many unexplored topics in haptic rendering of fine manipulation. Besides geometric features, fine features also include physical features, such as different textures and sharp changes of textures. For an object consisting of multiple parts with different physical properties, small movement of the tool in a time step may produce abrupt changes of the interaction force. For example, healthy tooth and decay have different stiffness. When a dental probe slides along the boundary of the two surfaces with different stiffness, there should be an abrupt change of force feeling. Haptic simulation of cutting an object involving different stiffness is also a challenging topic, which involves simulating topology change of the object while maintaining stability. For example, for a piece of beef steak including deformable meat and rigid bones, it is an open problem to simulate the subtle force/torque feeling when a knife cuts across both meat and bones.